Anatomy and Histology of the Pig

This comprehensive anatomy and histology book is the first guide specific to the pig since 1982. The high-quality, detailed full-color illustrations are accompanied by written material specific to each porcine anatomic section, with English-labeled terms for the various anatomic structures. The book covers all the major areas of interest in swine, including cardiovascular, digestive, neurologic, and thoracic structures. Differences between domestic and miniature pigs are discussed and a full-color histology section concludes the book.

The author team includes an experimental surgeon, a veterinary anatomist, and a veterinary pathologist, ensuring this atlas is appropriate for a wide range of professionals. It will be an invaluable resource for veterinarians, researchers, research technicians, and students for many years to come.

Anatomy and Histology of the Pig

A Concise Full-Color Atlas

Christian Alfaro
M. Michael Swindle
Kristi Helke

CRC Press
Taylor & Francis Group
Boca Raton London New York

CRC Press is an imprint of the
Taylor & Francis Group, an **informa** business

First edition published 2026
by CRC Press
2385 NW Executive Center Drive, Suite 320, Boca Raton, FL 33431

and by CRC Press
4 Park Square, Milton Park, Abingdon, Oxon, OX14 4RN

CRC Press is an imprint of Taylor & Francis Group, LLC

ISBN: 9780367266615 (hbk)
ISBN: 9781041061823 (pbk)
ISBN: 9780429294464 (ebk)

DOI: 10.1201/9780429294464

Typeset in Sabon LT Pro
by Evolution Design & Digital Ltd (Kent)

DEDICATION

This book is dedicated to my parents, Sandra and Johel;
my wife, Katarina; and to the memory of my dear niece, Alba.

Christian Alfaro

This book is dedicated to my grandchildren:
Seth, Claire, Emmett, and Mary

M. Michael Swindle

This book is dedicated to my parents,
Howard and Sharon

Kristi Helke

Contents

PREFACE

This book is meant to be a comprehensive atlas of porcine anatomy and histology. There are currently not any comprehensive textbooks on swine which include both anatomy and histology. I have been working with swine in research for >40 years and have been asked many times by researchers and veterinarians whether there was such a book available.

This book should be considered to be a companion book to *Swine in the Laboratory: Surgery, Anesthesia, Imaging, and Experimental Techniques*, Third Edition (2016, CRC Press). In all three editions, I have included anatomic descriptions and drawings to aid in performing the techniques used in experimental surgery. I relied heavily on my personal experience with performing surgical and other experimental techniques using swine as a primary translational research model.

The publisher of this textbook put our team together in order to produce a more comprehensive anatomic atlas which should be useful to both researchers and clinical veterinarians. The unique aspect of this book is that the team of authors includes an experimental surgeon, a veterinary anatomist, and a veterinary pathologist. The key issue was the recruitment of Dr. Alfaro-Rojas as the anatomic artist who committed substantial amounts of time to complete the anatomic atlas. Dr. Helke was my pathologist at the Medical University of South Carolina when I was the Chair of the Department of Comparative Medicine. She has substantial experience as a veterinary pathologist in collaborating with experimental protocols by myself and others who used swine in research. She is presently the Chair of that department.

We thank CRC Press (Taylor & Francis) for their commitment to the completion of this reference resource and their previous commitment to the publication of *Swine in the Laboratory*. We also thank the many collaborators I have worked with in the international arena which allowed me to develop the knowledge and skills to successfully perform many experimental protocols which led to the creation of this textbook. We especially thank our expert reviewers of the book: Dr. Susan W. Herring, Dr. Thimmasettapa Thippeswamy, and Mr. Wofgang Weber. Their credentials are listed in the Expert Reviewers section that follows. Professor Thippeswamy and Dr. Weber have kindly donated their reviewer fees to the Iowa State University Scholarships program, which this book is pleased to support. We would also like to thank Ms. Stephanie Lane for her help in translating Latin labels to English.

We hope that the book will be a useful resource for veterinarians, researchers, research technicians, and students for many years to come.

M. Michael Swindle

Authors

Christian Alfaro, DVM, MSI, is a Doctor of Veterinary Medicine with a background in scientific illustration. He completed his veterinary degree at the National University of Costa Rica and has dedicated his career to bridging biomedical sciences and visual arts through anatomical education.

Dr. Alfaro earned his Master's in Scientific Illustration from Maastricht University and Zuyd University of Applied Sciences in the Netherlands, where he focused on developing models for anatomical and surgical training. He has held academic and technical positions at the National University of Costa Rica and the Institute of Veterinary Anatomy at the University of Zurich, Switzerland. His work combines over a decade of experience in veterinary clinical practice, teaching, and the development of both physical and digital anatomical models.

He currently lives in Geneva, Switzerland, where he continues to create innovative anatomical resources that integrate accuracy, accessibility, and visual clarity to enhance the learning experience.

M. Michael Swindle, DVM, DACLAM, DECLAM (Retired), is a Professor Emeritus and retired Chairman of the Department of Comparative Medicine at the Medical University of South Carolina. He received his BS and DVM from Texas A&M University and is a Diplomate of the American College of Laboratory Animal Medicine and the European College of Laboratory Animal Medicine. He was in the US Army, ran a private veterinary practice, and was at Johns Hopkins Medical School. He has received several research awards and was honored as an outstanding alumnus of the College of Veterinary Medicine. His many publications and presentations are mainly in the areas of experimental surgery, anesthesia, and swine as animal models. He continues to be active in research and teaching as a private consultant for using porcine models.

Kristi Helke, DVM, PhD, DACVP is Professor and Chair of the Department of Comparative Medicine, MUSC College of Medicine. She received her Doctor of Veterinary Medicine degree in 2000 from the University of Wisconsin–Madison. Her PhD in pathobiology was earned at Johns Hopkins Medical School in 2007, where she also completed her pathology residency following a year of private veterinary practice in Goose Creek, SC. Dr. Helke has been a veterinary pathologist at MUSC since 2007. During this time, she has developed numerous collaborations with researchers in many fields of interest at MUSC, including neuroscience, lipidomics, cancer, and transplant research. She has several successful collaborations with local biotech companies, and she oversees their animal studies, which are performed at MUSC. Dr. Helke is a Diplomate of the American College of Veterinary Pathologists and has served the following professional organizations either as a member of the board of directors, committee chairman, and/or officer: Southeastern Association for Laboratory Animal Science, American College of Veterinary Pathologists, Society for Toxicologic Pathology, and the Trident Veterinary Medical Association. Her publications and presentations are mainly in the areas of neuroscience and cancer, as well as pathology, with many collaborative publications with researchers in varied fields.

EXPERT REVIEWERS

Susan W. Herring, PhD
Professor Emeritus, Orthodontics
Professor, Oral Health Sciences
School of Dentistry
University of Washington

Thimmasettappa Thippeswamy, DVM, MVSc, PhD, PgCert HE & TL
Professor
Department of Biomedical Sciences
College of Veterinary Medicine
Iowa State University

Wolfgang Weber
Teaching Laboratory Specialist II
Department of Biomedical Sciences
College of Veterinary Medicine
Iowa State University

SECTION 1
INTRODUCTION AND GENERAL ANATOMY

M. Michael Swindle, DVM, DACLAM, DECLAM (Retired)
Professor Emeritus, Department of Comparative Medicine
Medical University of South Carolina, Charleston, SC
Veterinary Consultant
International Swine Consulting
Marshall, NC

This book is divided into three sections: Anatomic Descriptions, Anatomic Atlas, and Histology. Much of the written information is derived from Swindle M.M., Smith A.C.: *Swine in the Laboratory: Surgery, Anesthesia, Imaging and Experimental Techniques,* Third Edition, Boca Raton: CRC Press (Taylor & Francis), 2016. The anatomic descriptions and drawings in that book are largely based upon 40 years of experience with using swine as a surgical model by Dr. Swindle and this book should be considered complementary to that book. The introductory material in each of the chapters in that book discusses comparisons to human anatomy as well as data on organ sizes, weight, and other physiologic characteristics. This section provides an overview of the general anatomy of swine (*Sus scrofa domesticus*), emphasizing the unique variations in swine anatomy.

The Anatomic Atlas section is largely based upon the dissections of porcine cadavers performed by Dr. Alfaro and the Histology section is based upon the pathologic studies of pigs used in research by Dr. Helke. The dissections and histology are based mainly upon 20–30 kg domestic research swine. However, miniature breeds have also been used by the three authors and when there are differences between miniature pigs and domestic swine, they are described in the Anatomic Descriptions section of the book. The decision to divide this book into three separate but complementary sections is based upon the authors' experiences with how they approach anatomy and histology when reviewing background information in order to conduct experimental procedures. However, the book should also be useful to students, agricultural workers, and others interested in some aspects of swine anatomy.

TOPOGRAPHICAL ANATOMY

The pig is a relatively hairless animal with a fixed skin tightly attached to the subcutaneous tissues, similar to that in humans. However, the skin of the pig is thicker and less vascular overall than human skin. The thickness of the skin is especially pronounced in sexually mature animals on the dorsal surface of the neck and back, and in some miniature breeds such as the Yucatan. There are also differences in the accessory tissues, such as the variations in sweat glands and the presence of an intrafollicular muscle in swine. Pigs have sebaceous glands, which are relatively insignificant. Eccrine glands in pigs are limited to the snout and carpal glands. Apocrine sweat glands are more extensive in pigs but do not contribute to sweating or thermoregulatory functions to a high degree. Secretions on the skin may help to prevent fluid loss, but pigs living or housed outdoors thermoregulate by wallowing or seeking shade when subjected to elevated temperatures, rather than actively sweating. There is a mental gland between the mandibles that contains tactile hairs and provides sebaceous and apocrine secretory functions. The snout (rostrum) contains two nostrils and blends with the upper lip. The upper lip has a notch at the level of the upper canine tooth which becomes the tusk when it projects laterally. Some breeds, such as the Yucatan, have a pair of wattles extending ventrally from the mandibular area.

HEAD AND NECK

There is a substantial difference in the conformation of the head and neck among different breeds. The snout of the Yucatan, Göttingen, Sinclair, and most other miniature pigs is considerably shorter than that of domestic farm breeds and the Hanford miniature pig. The heads of miniature breeds tend to be more rounded than those of the farm breeds and the Hanford; the latter has a head and snout shape similar to wild pigs.

Skull

The bones of the cranium and mandible are massive and the shapes and locations of the bones as well as the various foramina they contain can be best understood by studying the illustrations in the Atlas section. The occipital bone forms the caudal aspect of the skull and contains the foramen magnum. The pig has an elongated paracondyloid process extending from the ventral occipital bone. The sphenoid bone is located at the base of the cranium. The ethmoid bone is deep in the nasal cavity and forms turbinates separating the cranial and facial parts. The parietal bones are paired and join the frontal and occipital bones. The temporal bones form a major portion of the lateral wall of the cranial fossa. The vomer bone forms the ventral portion of the nasal septum. The maxillae are the bones of the upper jaw which contain the upper cheek teeth. The nasal bone is paired and forms the dorsum of the nose and internally a major part of the roof of the nasal cavity. The paired incisive bone is a rostral part of the bony skull and has labial and palatine surfaces. The palatine bones form the caudal portion of the hard palate. The pterygoid bones are caudal to and articulate with the palatine bones. The lacrimal bones form the medioventral portion of the bony orbit. The zygomatic bones articulate with the maxillae and lacrimal bones, extending the bony orbit ventrally and laterally.

The mandible is the lower jaw and the largest bone of the skull. The temporomandibular joint has been described as being similar to humans. The pig has a reciprocally fitting meniscotemporal joint and a condylomeniscal joint of the condylar type. The hyoid bone is not articulated with the other bones but provides the attachment of muscles of the tongue, the larynx, the pharynx, and the epiglottis.

The paranasal sinuses of the pig are extensive and consist of the maxillary, frontal, lacrimal, sphenoid, and conchal sinuses. The frontal sinus of the pig is important because it tends to enlarge and excavate the entire dorsal surface of the skull. Consequently, as the pig ages this results in two bony layers protecting the skull. This becomes important if cranial implants are implanted in immature animals because the implants will become encased in the sinuses over a few months.

Oral Cavity and Dental

The oral cavity conforms to the shape of the head of the particular breed. The tongue is long and thin and contains vallate papillae, numerous fungiform papillae, and small filiform papillae.

The dental formula for deciduous teeth in swine is 2(I 3/3, C 1/1, P 4/4) = 32. Permanent teeth have a dental formula of 2(I 3/3, C 1/1, P 4/4, M 3/3) = 44. Swine are born with the last incisors and canine teeth. There are slight differences among breeds and between miniature pigs and domestic swine. Some literature refers to the deciduous molars as deciduous premolars, and there can be a late eruption of M1 which can also be absent in miniature pigs, giving a deciduous formula of 2(I 3/3, C 1/1, M 3/3) = 28. Since the nomenclature of the molars and premolars varies in the literature, the formula can be confusing. The remainder of the incisors erupt between 2 weeks and 2 months of age. The premolars erupt between 4 days and 5 months. The molars are the first permanent teeth to erupt and appear between 4 and 20 months of age. The incisors turn over between 8 and 20 months, the canines between 9 and 10 months, and the premolars between 12 and 15 months. The tooth eruption sequence of adult permanent teeth is P1/M1/I3/C/M2/I1/P3/P4/P2/I2/M3. All teeth except the canines are brachydontic, whereas the canine teeth are hypselodontic and as such continuously grow. In the boar, the canine teeth become the tusks and require trimming two to four times a year in the adult. The dental formulas of adult swine are identical among farm and miniature breeds, and eruption dates are similar. A full set of permanent teeth is usually present by around 18 months of age.

The pharynx contains a pharyngeal diverticulum on its dorsal wall which is a blind pouch above the entrance to the esophagus. There are both paraepiglottic tonsils and a tonsil in the ventral surface of the pharynx.

The larynx is relatively elongated and contains lateral ventricles close to the entrance to the trachea as well as the following cartilaginous structures: epiglottic, arytenoid, thyroid, and cricoid cartilages. The vocal folds (ligaments) are prominent and extend dorsocaudally from the arytenoid to the cricoid cartilages. There is a cricothyroid membrane connecting the larynx to the trachea on the ventral surface.

Salivary Glands

The salivary glands of the pig are the parotid, the mandibular, and the sublingual. The parotid gland is located at the base of the ear and enters the oral cavity opposite the upper fourth premolar or first molar and may have accessory glands along the duct. The mandibular gland is located medial to the ramus of the mandible and the duct enters the oral cavity near the frenulum. The sublingual glands are located in bilateral chains in the floor of the mouth and have multiple openings into the mouth. A series of minor buccal glands are located opposite the upper and lower cheek teeth. In the pig, the parotid gland is serous, the sublingual gland is mucoid, and the mandibular and buccal glands are mixed in secretions.

Eyes

The eyes are deeply embedded in the orbits. They are considered to have binocular vision with some peripheral vision and possibly some color vision. The eyelids are thickened and contain fatty tissue and glandular structures. They have a third eyelid in the ventral portion of the orbit which has both a deep (Harderian) gland and a superficial (nictitating) gland. The lacrimal gland is serous and located on the dorsolateral aspect of the eye.

The seven bulbar muscles are the dorsal rectus, ventral rectus, medial rectus, lateral rectus, dorsal oblique, ventral oblique, and retractor bulbi. The retractor muscle allows the eyeball to be retracted into the orbit which allows the third eyelid to cover and protect the eye.

The cornea is horizontally oval in shape and is thicker at the dorsal and ventral limbic area than in the center. Corneal thickness measurements vary widely depending upon the breed and age of the animal as well as the methodology and area of measurement. A true Bowman's membrane is absent. Differences in corneal thickness between minipigs and farm pigs are likely because of the differences in body size.

Instead of a single canal of Schlemm, there are multiple small channels making up part of a scleral venous plexus. The sclera and iris are usually pigmented.

The retina lacks a tapetum cellulosum and foveolar specialization. Pigs do not have a tapetum lucidum; consequently they do not reflect light as some other species do. The retinal vascular pattern is holangiotic with the retinal vessels arising from central arteries. The optic nerve passes through the lamina cribrosa. Overall, the porcine retina is comparable to that of the human.

Auditory System

Pigs have a highly developed auditory system. The external ear varies considerably in size and shape among breeds. In some breeds it is erect and in others the pinnae hang down over the face (lop ear). The shape of the ear is determined by the auricular cartilage. The internal anatomy of the ear contains the structures common to other species of mammals but may vary considerably in size.

Olfactory System

The olfactory bulbs are large, as would be consistent with the heightened sense of smell reported in the pig. The nasal cavity contains two nasal conchae which divide the nasal cavity into three meatuses. Ethmoidal conchae contain the extensive olfactory mucosa.

Thyroid and Parathyroid Glands

The thyroid gland is located on the ventral surface of the trachea and, unlike other species, is caudally distant from the larynx and is located near the thoracic inlet. The lobes of the gland are fused and appear as a single gland along the ventral trachea. The small paired parathyroid glands are located in the cranial portion of the thymus. The paired thymus is extensive and extends from the larynx to the base of the heart on either side of the trachea in the young pig. As the pig matures the thymus in the neck shrinks and migrates caudally towards the thorax.

MUSCULOSKELETAL SYSTEM

The skull is described above in the section on the head and neck.

The typical vertebral formula is C7, T14–15, L6–7, S4-5, Cy20–23. Some miniature breeds may have one less thoracic or lumbar vertebra or one less of both, which would alter the vertebral formula above. There are seven sternal and seven asternal ribs. If a 15th rib is present, it is usually a floating rib rather than one attached to the cartilage of the costal arch. The clavicle is absent.

In the forelimb, there are eight carpal bones, four metacarpal bones (2–5), and three phalanges per digit; proximal and distal sesamoid bones are present. In the hindlimb, there are eight tarsal bones, four metatarsal bones, and phalanges with sesamoid bones present.

The musculature of the pig is massive, as would be expected of an animal that has been bred predominantly for meat production. Their muscles have a predominance of Type II B muscle fibers with a lesser extent of Type IIA, IIC, and IIX fibers. Bone can be subdivided into two structural types: cancellous, also known as trabecular, spongy, or porous bone, and cortical, also known as compact bone. This classification is based on porosity and unit microstructure.

Hyaline cartilage covers bone and joint surfaces. Fibrocartilage is found in intervertebral discs, symphysis pubis, and certain joints. Elastic cartilage is found in the auditory tubes, external ear, epiglottis, and larynx.

Skeletal maturity, which is defined as closure of the epiphyses of the long bones, varies among breeds of swine. Dates of epiphyseal closure for some of the breeds of miniature pigs are: Hanford minipigs, 2–4 years; Yucatan minipig, 2–3 years; Sinclair minipig, 2.5–3 years; Göttingen minipig, 1.5–3 years; and Yucatan micropig, 1.5–2 years. Domestic breeds have complete epiphyseal closure at between 3 and 4 years depending upon the breed. These data are a guideline but are not precise due to genetic, gender, and breed variances.

The muscles and bones are best studied using the illustrations in the Atlas section of this book. Unique features of them are noted in the text which follows.

Head and Neck

The masseter muscle is the thickest muscle of the head. The musculature of the facial region, the hyoid, and the ear are similar to other species. The facial muscles which move the rostrum are large and specialized. The muscles of the ear will have some variations among breeds. The muscles of the eye are discussed above with the anatomy of the eye.

No ligamentum nuchae has developed in the pig since the short neck of the pig does not require it for support. The cutaneous colli is the most superficial muscle beneath the skin and subcutaneous tissues.

The other muscles of the neck are thickened and have origins and insertions typical of other species.

Forelimb (Thoracic Limb)

Pigs are ungulates, which means they ambulate on hooves. Pigs have an even number of hooves on each leg and are subclassified as artiodactyl, as opposed to animals with one hoof per leg. The principal or primary digits are numbers 3 and 4. The dewclaws or accessory hooves are non-weight-bearing accessory digits and are numbers 2 and 5. There are proximal, middle, and distal phalanges in the primary and accessory hooves. Sesamoid bones are located on the palmar surface of digits 3 and 4 at the joints of the middle and distal phalanges. There is also a single sesamoid bone between the first carpal bone and digits 2 and 3.

Muscles of the forelimbs are similar to those of other quadruped animals.

Hindlimb

The hindlimb of the pig has the typical bones and muscles of other quadrupeds. The digits and phalanges are numbered in the same manner as the forelimb. There is a sesamoid bone on the proximal caudal surface of metatarsal bone 3.

Thorax

The thorax can have 13–17 pairs of ribs depending upon the breed. Most pigs will have 14–15 pairs with the first seven being attached to the sternum. If there are more than 14 ribs the caudal ones may be floating without a costochondral component. The ribs are wide with narrow intercostal spaces.

There are six segments comprising the sternum. The manubrium is elongated with a small cartilaginous portion on its cranial tip. The xiphoid cartilage is at the caudal end. The diaphragm extends from the region of the xiphoid process dorsally to the caudal region of the rib cage. It contains a central tendinous center with lateral crura.

Abdomen

The abdominal wall extends from the diaphragm to the pelvis and contains similar muscle layers to those found in other quadrupeds. The major muscles are the cutaneous trunci, external abdominal oblique, internal abdominal oblique, rectus abdominis, and transverse abdominis. The musculature is relatively thick in most of the abdominal wall but thins as it approaches the attachment to the pelvis.

Pelvis, Sacrum, and Tail

The pelvic bones are dense and massive with the usual bony structures found in other mammals. The wings of the ileum are prominent and can be readily visualized beneath the skin in the intact animal. The sacrum consists of 4–5 vertebrae which become fused with age. The coccygeal vertebrae are variable in number (20–23) and the tail becomes curved after the first few vertebrae.

GASTROINTESTINAL TRACT

Unique features of the gastrointestinal tract include the torus pyloricus in the pyloric region of the stomach, the mesenteric vascular arcades, and the spiral colon. The torus pyloricus is a muscular and mucoid glandular structure adjacent to the pylorus, which is involved in the functional closure of the orifice. The small intestine is arranged in a series of coils from the stomach to the pelvis, dorsal and lateral to the spiral colon on the right side of the abdomen. The bile duct and pancreatic ducts enter the duodenum separately in the proximal portion in the right upper quadrant of the abdomen. The mesenteric vascular arcades form in the subserosa of the intestine rather than the mesentery, giving a fanlike appearance. The spiral colon contains the cecum, ascending colon, transverse colon, and main portion of the descending colon arranged in a series of centrifugal and centripetal coils in the left upper quadrant of the abdomen, caudal and ventral to the stomach. The descending colon continues caudally on the dorsal left side to become the rectum.

The esophagus originates from the pharynx at the second cervical vertebra and is dorsal to the trachea, becoming more left lateral in the caudal neck and thorax. The tunica muscularis of the porcine esophagus is composed of inner circular and outer longitudinal layers which are composed of striated muscle fibers in the proximal third, mixed striated and smooth muscle fibers in the middle third, and smooth muscle fibers in the caudal third of the esophagus. The mucosa is stratified squamous epithelium throughout the length.

The stomach is divided into esophageal, cardiac, fundic, and pyloric regions with a diverticulum in the fundic region. The secretory functions and glandular mucosa of these gastric regions are similar to those of other mammals.

Generally, the small intestine is 20 to 40 times the length of the pig. The duodenum generally is located in the right cranial medial aspect of the abdominal cavity in the region of the 10th–12th intercostal spaces. It is tightly attached and relatively fixed in its position in its cranial aspect, and the pancreas is closely adherent to it. The bile duct enters the duodenum approximately 1.5–6 cm distal to the pylorus, and the separate pancreatic duct enters the duodenum 10–20 cm from the pylorus. The distances increase with the increasing size of the pig; the smaller value represents weanlings, and the larger value represents adults. The duodenal portion of the small intestine courses medially and passes through the ligament of Treitz caudal to the stomach, after which it gradually develops into the jejunum. The duodenum is generally considered to be approximately the proximal 5% of the small intestinal mass.

The jejunum comprises approximately 90% of the small intestinal mass. Most of it is located in the right caudal ventral side of the abdominal cavity in tight coils. The gross differences between the jejunum and ileum are not distinct; however, the wall of the ileum is slightly thinner. The ileum is the distal 5% of the small intestine and generally originates in the middle of the abdominal cavity, in which it courses cranially to attach to the cecum in the spiral colon at the level of the left paralumbar fossa.

Aggregated lymph nodules (Peyer's patches) are found throughout the jejunum and ileum. A distinct aggregation is located at the ileocecal junction and extends for some variable distance into the cecum. The mesentery of the small intestine is thin and friable with a unique fanlike pattern of mesenteric vessels with vascular arcades forming in the submucosa. Prominent lymph nodes are present in the root of the mesentery.

The cecum contains three muscular longitudinal bands (teniae) and three corresponding rows of sacculations (haustra). The ascending colon continues clockwise from the cecum in thee spirals in the mesentery to the apex of the cone of the spiral colon on the ventrum of the abdomen. The clockwise outer coils of the ascending colon contain two teniae and two haustrae. The inner centripetal coils do not contain teniae or haustra. The colon then reverses itself inside the cone in a series of counterclockwise turns towards the dorsum where it continues as the transverse colon on the right side of the root of the mesentery. The descending colon passes caudally and is continued as the rectum in the pelvic cavity. The rectum continues caudally to the anus. The cecum, colon, rectum, and anus have an outer longitudinal muscle and an inner circular muscle layer. The levator ani muscle arises from the sacrosciatic ligament.

LIVER

The liver contains six lobes: the left lateral, the left medial, the quadrate, the right medial, the right lateral, and the caudate, which contains a caudate process partially surrounding the caudal vena cava. The gallbladder is located in a fossa formed by the left and right medial and quadrate lobes in the right upper quadrant of the abdomen. The hepatic bile duct exits the liver caudally dorsal to the gallbladder. It joins the cystic and common bile ducts shortly after entering the porta hepatis (portal fissure). The bile duct is located in the porta hepatis ventral to the portal vein and hepatic artery. It may be identified as a translucent tubular structure within the mesenteric attachments in this area, and it courses caudally to enter the duodenum approximately 2–5 cm from the pylorus as a separate duct from the pancreatic duct. The sizes of the bile duct and sphincter of Oddi at the entrance of the bile duct into the duodenum are variable depending upon the size and breed of the pig. The diameter of the bile duct varies substantially even in the same size and breed of pig.

The hepatic artery and the portal vein enter the liver in close approximation to where the common bile duct exits. Lymph nodes and fascia in the area make the dissection of these structures difficult. The liver of the pig contains fibrous septations between lobules, which are indistinct grossly but can be readily identified histologically.

PANCREAS

The pancreas of the pig is extensive, and the tail follows the lesser curvature of the stomach from the spleen and left kidney to a position along the proximal duodenum. The majority of the pancreas is retroperitoneal. The head of the pancreas (duodenal lobe) gives rise to the body and tail. The pancreas encircles the portal vein and superior mesenteric artery and extends dorsally to the region of the left kidney. The pancreatic duct is composed of two separate ducts draining the tail and body. They anastomose to form the common pancreatic duct (accessory duct) immediately prior to the pancreatic sphincter. The pancreatic duct enters the duodenum caudal to and separate from the bile duct in the proximal duodenum approximately 20–25 cm distal to the pylorus in the area of the first curve of the duodenum. The duct is a firm, whitish structure along the caudal third of the portion of the pancreas that is associated with the duodenum. There is usually a single pancreatic artery supplying the tail (splenic lobe) of the pancreas as a branch of the splenic or common hepatic (gastrohepatic) artery. The pancreatoduodenal artery courses between the duodenum and the pancreas along its joint border and supplies both from a series of small branches. Venous drainage is through the splenic vein. Some variations in vascular and ductal anatomy may be encountered among individuals and breeds.

SPLEEN

The spleen is a pedunculated organ that is elongated in shape and located in close apposition to the greater curvature of the stomach in the left upper quadrant of the abdomen. It extends from the left kidney ventrally to the midline. There are three main vascular supplies to the organ. These are located in the splenic ligament and are the left gastroepiploic, the splenic, and the short gastric arteries and veins. The vascular supply enters the organ from the head to one half the distance to the tail of the spleen.

ADRENAL GLANDS

The adrenal glands are craniomedial to the hilum of the kidneys. The right gland is attached to the vena cava. The arterial supply of the glands is from either the aorta or branches of the lumbar arteries and the venous drainage is into either the vena cava or the renal veins.

URINARY SYSTEM

The kidneys are usually ventral to the transverse processes of the first four lumbar vertebrae with the left kidney usually more cranial than the right one.

The cranial pole of the left kidney is located at approximately the 13th rib (L1–L4). The renal artery and vein divide into two branches close to the renal hilum. The blood supply is divided into cranial and caudal segments rather than into longitudinal halves as in other species. This means that the avascular plane of the kidney is transverse rather than longitudinal. The multipyramidal kidney has an internal renal anatomy similar to humans. The kidney has two major calices and a variable number of minor calices emptying into them.

The ureters extend caudoventrally from the renal hilum to the dorsolateral aspect of the neck of the urinary bladder. The urinary bladder is large and thin-walled but typical in morphology to other species. It receives its innervation from spinal nerves S2 to S4. The male urethra courses along the pelvic floor into the penis. The tip of the penis is located on the ventral abdominal wall in a preputial diverticulum. The female urethra enters the ventral floor of the vagina within the pelvic cavity, cranial to the vestibule. There is a ventral urethral diverticulum at the entrance of the urethra into the vagina.

REPRODUCTIVE TRACT

Female

The reproductive tract of the female is typical of that of a bicornuate species that produces litters. The ovaries are located caudal to the kidneys and are only loosely attached by a thin ovarian ligament and suspended within the broad ligament of the uterus (mesometrium). The ovarian vessels are the last branches of the aorta and vena cava prior to the iliac bifurcation at approximately L5–L6. The fallopian tubes are long and tortuous and typically form coils in the caudal abdominal cavity. The uterine horns are long and curve cranially from the ovaries and then reverse direction to form the short body of the uterus at approximately the same region as the ovaries in the midsagittal plane. The cervix is thick, elongated, and has a curved cervical canal containing rows of mucosal prominences. The vagina extends caudally in the pelvic cavity and contains the urethral orifice on the ventral floor at approximately the level of the caudal edge of the pubic bone. The pig has a diffuse epitheliochorial placentation. This type of placenta does not invade the endometrium. The placental membranes include the yolk sac, amnion, allantois, and chorion, and the latter two membranes fuse at an early stage to form a chorioallantoic type of placenta. The chorioallentois is responsible for transplacental transport of nutrients from the sow.

Paired mammary glands are located on the ventral surface of the abdomen and extend from the caudal thorax to the caudal inguinal regions. The number of glands is highly variable with an average of 6–7 pairs. Generally, breeds that produce large litters have a greater number. The vasculature is located along the lateral edges of the row of glands. The pectoral and cranial abdominal glands receive their blood supply from branches of the internal thoracic artery, which continues along the glands on the abdomen as the cranial superficial epigastric artery. Blood from the mammary glands drains into the cranial superficial epigastric vein, which continues cranially as the internal thoracic vein. The caudal glands (2–3 pairs) receive their blood supply from the external pudendal arteries, and drainage is from the pudendal veins into the iliac system.

Male

The male reproductive system is typical of domestic animals in location; however, there are several variations in anatomy that are important. The scrotum and testicles are located in the perineal region and may be of considerable size in adult males. The spermatic cord passes through each inguinal ring in the caudal abdomen. The spermatic vessels branch off the aorta and vena cava just cranial to the iliac bifurcation. Each ductus deferens enters the urethra independently on the dorsal surface at the neck of the bladder. The accessory sex glands are similar to those of humans except for the prominence of the various structures. The paired vesicular glands are large and located on either side along the neck of the bladder. The prostate is small and located on the dorsal surface of the urethra at the entrance of the ductus deferens. Paired bulbourethral glands extend along the dorsolateral surface of the urethra starting at the caudal brim of the pubis; these may extend along the entire length of the pubis in intact adults. The crus of the penis and the ischiocavernosus muscles form at the caudal surface of the pubis with the base of the penis. The penis extends ventrally and cranially forming a fibromuscular sigmoid flexure as it curves from the perineum to the ventral surface of the abdomen. The penis extends almost to the umbilicus before terminating in the preputial diverticulum in a corkscrew shape.

RESPIRATORY SYSTEM

The snout and nares are described above in the head and neck section. The larynx is also described more completely in that section. The larynx is prominent with a large vestibule that narrows caudally because of internal compression of the cricoid cartilage. There are middle and lateral ventricles that must be avoided during intubation.

The trachea extends from C4–5 to T5 where it bifurcates into the bronchi. The dorsum of the trachea has the right side of the cartilage ring slightly overlapping the left side.

At T3, the trachea provides a unique apical lobe bronchus to the apical lobe of the right lung. The lungs are composed of apical, middle, and diaphragmatic lobes, with an accessory lobe to the right lung. The intralobular fissure is incomplete between the left apical and middle lobes.

CARDIOVASCULAR SYSTEM

The heart extends from T2 to T7 and has sternal contact over most of the caudal distance. The left azygos (hemiazygos) vein is the most unique feature of the porcine vascular anatomy. It curves caudoventrally from the dorsal thorax across the dorsal surface of the heart ventral to the left hilum of the lung to enter the right atrium in the coronary sinus.

The coronary circulation of the pig has few subepicardial collateral anastomoses, like 90% of the human population. The circulation to the conduction system is predominantly right-side dominant from the posterior septal artery.

There are also differences from humans in physiological composition of the conduction system. The pig endocardium and epicardium are activated simultaneously because of differences in distribution of the specialized conduction system in the ventricles. The aorta has a true vasa vasorum, unlike many species.

Peripheral veins in the pig have valves like humans, and this may be a problem with passage of catheters into the small vessels in the extremities.

NERVOUS SYSTEM

Comparisons in morphology and function of the porcine central nervous system (CNS) to humans have been made. Briefly, the similarities to humans are as follows: gyrencephalic morphology, distinct caudate and putamen structures in the striatal portion, cytoarchitectonic similarities in the cortex and hippocampus, and some dopaminergic nuclei functions. In general, the patterns of distribution of gray and white matter, cerebral blood flow parameters, the size, and brain growth during development are similar. The spinal cord terminates with the conus medullaris at S2–S3 and follows down the remaining vertebrae with the cauda equina and the filum terminale.

The common carotid arteries leave the brachiocephalic trunk at the level of T1. The internal carotid artery provides most of the circulation to the brain. Vertebral arteries branch off the subclavian arteries at the level of C6–C7 and supply the vertebrae, muscle, and spinal cord at each level. The cranial portion of the vertebral artery provides some blood supply to the medulla as well. In studies that interrupted the blood supply of the branches of the vertebral artery at C1–C2, C4–C5, and C7–T1, the latter produced the fewest hemodynamic and systemic changes. Venous plexuses from the brain enter the internal jugular vein at the level of C1–C2. The internal vertebral plexus enters the vertebral veins and drains into the azygos system at the level of C6. Venous drainage from the thoracic vertebrae enters the azygos veins cranial to the heart.

Retia mirabilia are mesh-like vascular structures which are located at the base of the skull in cavernous sinuses. They connect the ascending pharyngeal arteries and internal carotid arteries in a communication across the midline. The size of the vascular structures within the net are small, approximately 100 μm in the size of pigs commonly used in research. However, they will vary in size consistent with the size and maturity of the pig. The retia mirabilia are an obstruction to catherization of the cerebral vessels but do provide a net of protection from emboli.

The cranial nerves I–XII are similar in their anatomic location and function as in other mammalian species. Primary innervation areas of the nerves and their branches are as follows:

- I – Olfactory: nasal cavity
- II – Optic: retina
- III – Oculomotor: extrinsic and intrinsic muscles of the eye
- IV – Trochlear: superficial oblique muscle involved in eye movement
- V – Trigeminal: sensory and motor functions of the facial structures. Major branches are ophthalmic, maxillary, and mandibular nerves
- VI – Abducens: muscles which provide lateral movement of the eyeball
- VII – Facial: muscles of facial expression and movement
- VIII – Vestibulocochlear: vestibular and cochlear functions of the inner ear
- IX – Glossopharyngeal: sensory and motor functions of the pharynx, caudal tongue, tympanic membrane, and carotid sinus
- X – Vagus: sensory and motor functions, pharynx, larynx, recurrent laryngeal nerves, joins the vagosympathetic trunk to the heart and lungs, forms dorsal and ventral trunks along the caudal esophagus, stomach, liver, pancreas, and intestines providing parasympathetic functions of the viscera
- XI – Accessory: runs in association with the vagus upon exiting the jugular foramen. It divides into internal and external branches at the ascending pharyngeal artery and then the internal branch joins the ganglion with the vagus. The external branch divides into a dorsal branch which interacts with cervical nerves II, III, and IV. The ventral branch controls neck and trunk muscles which allow head movement and rotation (sternocephalicus, brachiocephalicus, trapezius)
- XII – Hypoglossal: motor function of the tongue and associated muscles allowing functions such as swallowing

Spinal Nerves

Spinal nerves are divided into cervical, thoracic, lumbar, sacral, and coccygeal sections. The numbers of nerves in each section is dependent upon the number of vertebrae. They provide motor, sensory, and autonomic functions. Dorsal branches emerge from the lateral vertebral foramina and ventral branches from the intervertebral or ventral lateral foramina. Dorsal roots are afferent nerves which conduct impulses from the body to the spinal cord and brain. Dorsal rami are mixed. Dorsal and ventral roots join to form the mixed spinal nerve. Efferent nerve cell bodies are located in the ventral horn and conduct impulses from the brain and spinal cord to the body.

Cervical nerves (CNs) form eight pairs which divide into medial (dorsal) and lateral (ventral) branches. The dorsal branches terminate as cutaneous nerves or motor branches. The ventral branch of CN1 joins the hypoglossal nerve and then leaves it to form the superior limb of the ansa cervicalis. The cervical plexus forms the rest of the ansa. Ventral branches of CN2–5 innervate the lateral and ventral cutaneous regions of the neck. Ventral branches of CN5–7 combine to form the phrenic nerve which controls the diaphragm.

The number of thoracic nerves depends upon the number of thoracic vertebrae, which is between 13 and 17 depending upon the breed. The branching of the thoracic nerves is similar to the cervical nerves except the ventral branches emerge through ventral lateral foramina instead of intervertebral foramina. They innervate cutaneous and muscular regions of the thoracic wall.

Lumbar nerves emerge from the lumbar vertebrae, which are usually 5–6 in number, in a similar fashion to the thoracic nerves.

Sacral nerves are usually four in number. They provide innervation of the structures and organs of the pelvic region.

Coccygeal (caudal) nerves vary in number, usually 4–8 pairs, and innervate the structures of the tail.

A plexus is formed by specific spinal nerves as they course to the limbs. The brachial plexus is formed by ventral branches of cervical nerves 5–8 and the first thoracic nerve. Its primary function is innervation of the forelimb. Major nerves of the brachial plexus are the suprascapular, subscapular, pectoral, musculocutaneous, axillary, radial, ulnar, median, long thoracic, thoracodorsal, and lateral thoracic.

The lumbar plexus is formed by ventral branches of the lumbar spinal nerves. This plexus innervates the paralumbar region of the abdominal wall, the mammary and inguinal areas, and the hindlimb. Major nerves of this plexus are the femoral, saphenous, and iliohypogastric nerves and their branches.

The sacral plexus follows the medial wall of the pelvis and mainly innervates the hindlimb and pelvic region. Major nerves include the lumbosacral trunk, gluteal nerves, pudendal, caudal cutaneous femoral, sciatic, ischiatic, fibular, and tibial nerves.

The autonomic nervous system is the part of the peripheral nervous system which consists of sympathetic and parasympathetic divisions. The autonomic system regulates and controls involuntary functions such as heart rate, blood flow, respiration, urinary, and digestive functions, mainly through the parasympathetic functions. The sympathetic system is associated with more stimulatory functions such as fight or flight syndromes, but is also important in homeostasis.

The sympathetic division of the autonomic nervous system may be divided into cervical, thoracic, abdominal, pelvic, and caudal parts. Sympathetic ganglia are formed on either side of the vertebral column.

The cervical sympathetic trunk extends from the cranial cervical ganglion to the caudal cervical ganglion passing through a series of three vertebral ganglia. Spinal nerves C2–8 from these ganglia have their effects on structures and vessels of the head and neck as well as the forelimbs and cranial thorax.

The thoracic part of the autonomic system continues from the cervical part. It extends from the cervicothoracic ganglion near the first rib along the vertebrae into the abdominal cavity. Thoracic sympathetic trunk ganglia are present in the intercostal spaces. Spinal nerves T1–L2 carry sympathetic innervation for the heart, lungs, blood vessels, and thoracic wall. After the thoracic sympathetic trunk passes into the abdominal cavity, the abdominal viscera are innervated. In the abdominal cavity the lumbar and caudal portions provide sympathetic innervation to the viscera of the abdominal and pelvic cavities.

The parasympathetic nervous system is predominately composed of the vagus nerve (CN X) and its branches. The vagus nerve originates in the CNS and exits the calvarium in the petro-occipital fissure. It progresses though the neck in the jugular furrow into the thorax and cranial abdomen following the esophagus. Primary branches of the vagus nerve are the recurrent laryngeal, cardiac, bronchial, and esophageal nerves. It passes through the diaphragm and follows the aorta with branches providing parasympathetic function to the stomach, pancreas, liver, kidneys, and intestines. Sacral nerves provide parasympathetic functions to organs and structures in the pelvic cavity. Parasympathetic function of the innervation of the structures of the head originates from other preganglionic segments of other cranial nerves. The recurrent laryngeal nerves do not follow the course of the vagus but reverse direction to innervate the larynx in the same sheath as motor and sensory nerves.

LYMPHOID SYSTEM

The regional distribution of the lymphatic system of the pig is illustrated in the Atlas in this book. Histologically the lymph nodes of the pig have an inversion of the cortex and medulla as compared to other mammals. The majority of the germinal centers are located centrally rather than in the cortical position. A histologic section of a porcine lymph node is illustrated in the Histology section of this book.

Numerous lymph nodes are located throughout the body in subcutaneous and deep tissues. Lymph nodes are located throughout the thoracic, abdominal, and pelvic cavities in association with the various organs and tissues in those areas.

Tonsils are located in the pharyngeal region of the oral cavity. A paraepiglottic tonsil is located at the base of the epiglottis. Other tonsils are located in the soft palate, the roof of the pharynx, and adjacent to the auditory tube.

The lumbar trunk along with the various trunks from the abdominal viscera form the abdominal cisterna chyli which is located dorsolateral to the abdominal aorta in the vicinity of the renal arteries. The thoracic duct continues from the cisterna chyli along the aorta and is located lateral to the aorta on the right side of the thorax. It drains into the cranial vena cava.

The spleen is discussed with the abdominal organs and the thymus is discussed with the thyroid and parathyroid glands in the head and neck section.

GENERAL REFERENCES

Sack, W. O. 1982. Essentials of Pig Anatomy, Ithaca, NY: Veterinary Textbooks.

Schantz, L. D., K. Laber-Laird, S. Bingel, and M. M. Swindle. 1995. Pigs: Applied anatomy of the gastrointestinal tract. In Essentials of Experimental Surgery: Gastroenterology, edited by S. L. Jensen, H. Gregersen, F. Moody, and M. H. Shokouh-Amiri. NY: Harwood Academic Publishers, pp. 2611–9.

Sisson, S., W. P. Switzer, S. S. Gandhi, et al. 1975. Porcine. In Sisson and Grossman's The Anatomy of the Domestic Animals, 5th Edition, Volume 2, edited by R. Getty. Philadelphia, PA: WB Saunders, pp. 1215–424.

Swindle, M. M., and A. C. Smith. 1998. Comparative anatomy and physiology of the pig. *Scand J Lab Anim Sci*, 25 Suppl 1:1–10.

Swindle, M. M., and A. C. Smith. (Eds). 2016. Swine in the Laboratory: Surgery, Anesthesia, Imaging and Experimental Techniques, 3rd Edition. Boca Raton, FL: CRC Press (Taylor & Francis), 2016.

SECTION 2
ANATOMIC ATLAS

Christian Alfaro, DVM, MSI
Doctor of Veterinary Medicine
University of Costa Rica
Master's in Scientific Illustration
Maastricht University and
Zuyd University of Applied Sciences
The Netherlands

Skeleton of pig – lateral view

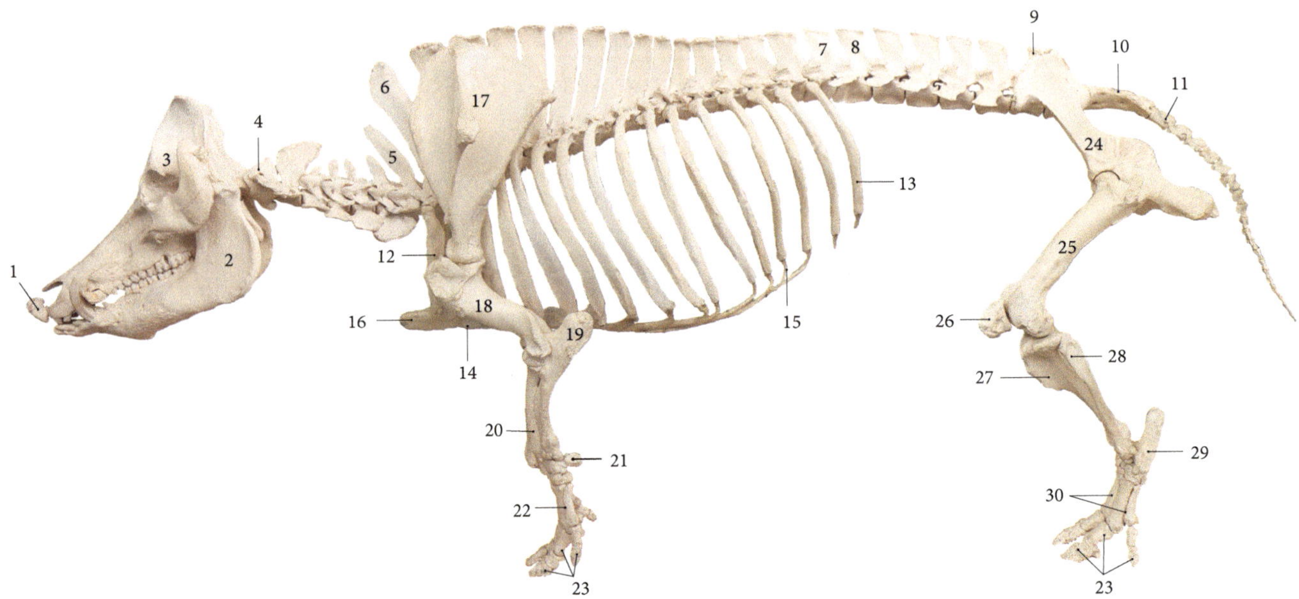

1. Rostral bone
2. Mandible
3. Skull
4. First cervical vertebra
5. Seventh cervical vertebra
6. First thoracic vertebra
7. 15th thoracic vertebra
8. First lumbar vertebra
9. Sixth lumbar vertebra
10. Sacrum
11. First caudal vertebra
12. First rib
13. Last rib
14. Sternum
15. Costal cartilage
16. Cartilage of manubrium
17. Scapula
18. Humerus
19. Ulna
20. Radius
21. Carpal bones
22. Metacarpal bones
23. Phalanges
24. Coxal bone
25. Femur
26. Patella
27. Tibia
28. Fibula
29. Tarsal bones
30. Metatarsal bones

Skeleton of pig – dorsal view

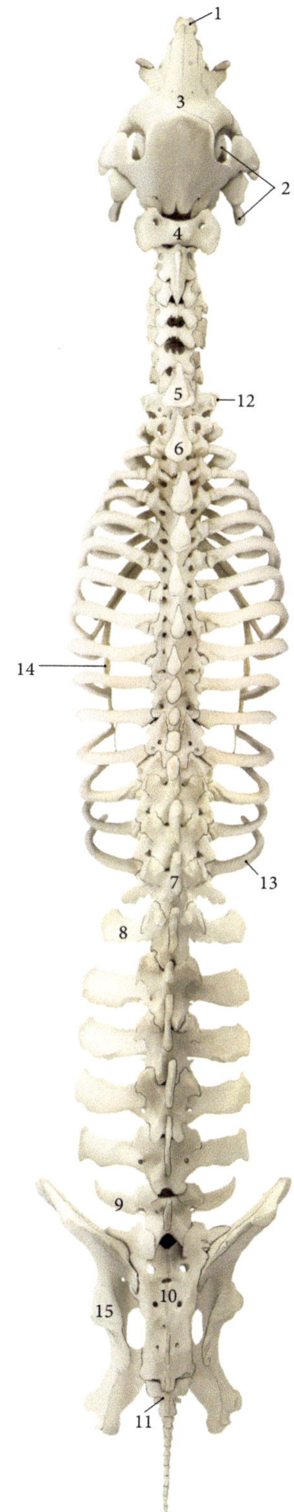

1. Rostral bone
2. Mandible
3. Skull
4. First cervical vertebra
5. Seventh cervical vertebra
6. First thoracic vertebra
7. 15th thoracic vertebra
8. First lumbar vertebra
9. Sixth lumbar vertebra
10. Sacrum
11. First caudal vertebra
12. First rib
13. Last rib
14. Costal cartilage
15. Hip bone

Skull of pig – lateral view

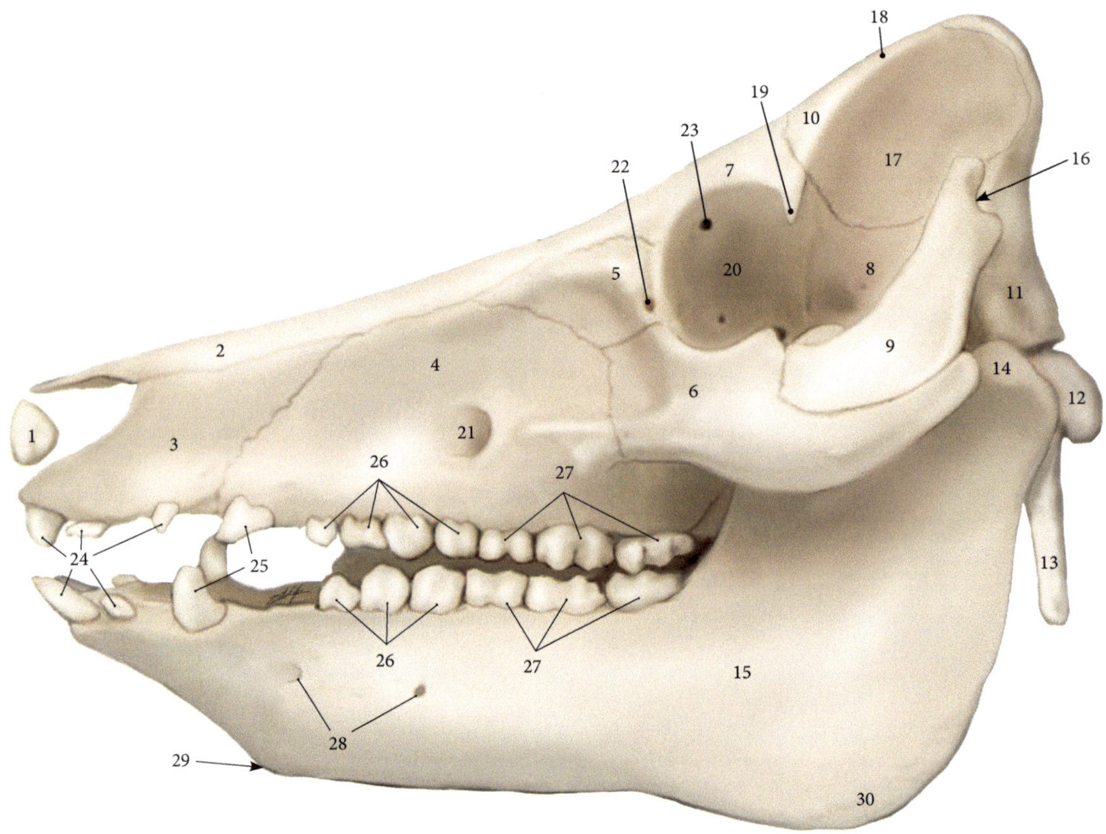

1. Rostral bone
2. Nasal bone
3. Incisive bone
4. Maxilla
5. Lacrimal bone
6. Zygomatic bone
7. Frontal bone
8. Squamous part of temporal bone
9. Zygomatic process of temporal bone
10. Parietal bone
11. Occipital bone
12. Occipital condyle
13. Jugular process
14. Condylar process of mandible
15. Mandible
16. External acoustic meatus
17. Temporal fossa
18. Parietal crest
19. Zygomatic process of frontal bone
20. Orbital part of frontal bone
21. Infraorbital foramen
22. Lacrimal foramina
23. Orbital opening of supraorbital canal
24. Incisor teeth
25. Canine teeth
26. Premolars
27. Molars
28. Mental foramina
29. Mental prominence
30. Angle of mandible

Skull of pig – sagittal view

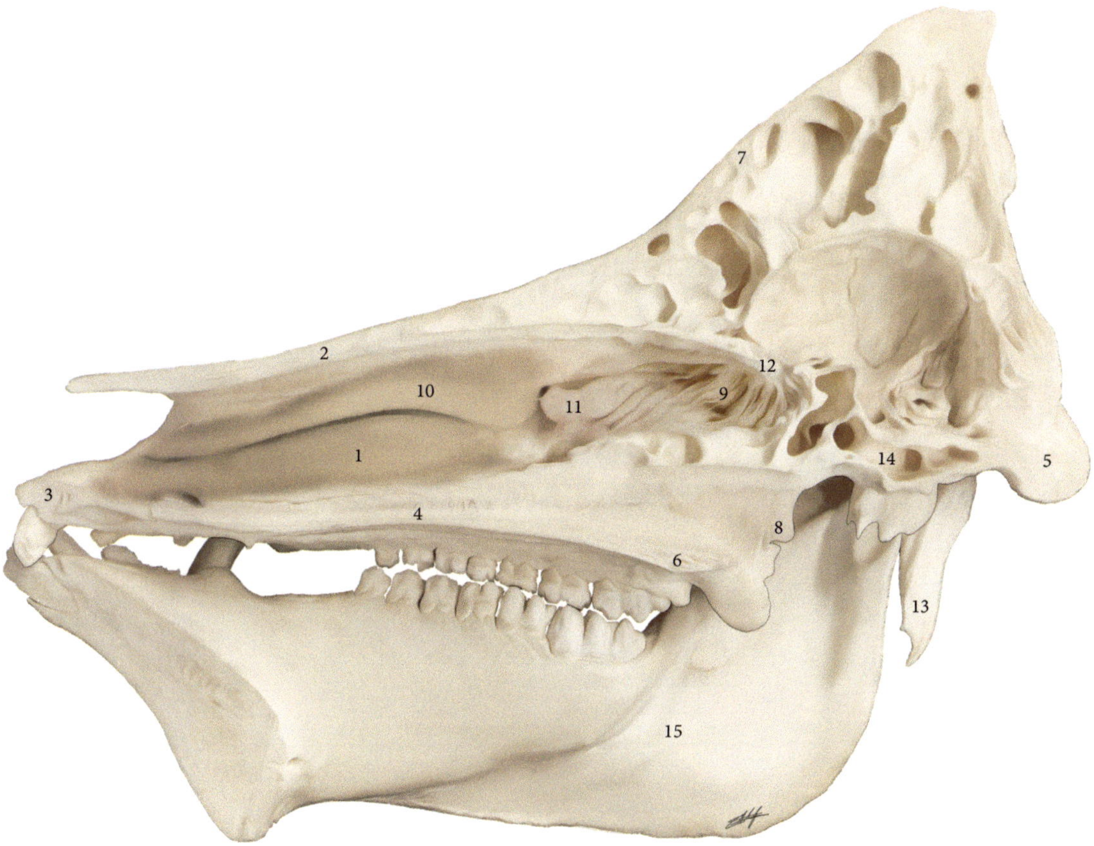

1. Ventral nasal concha
2. Nasal bone
3. Incisive bone
4. Maxilla
5. Occipital bone
6. Palatine bone
7. Frontal bone
8. Pterygoid bone
9. Ethmoturbinates
10. Dorsal nasal concha
11. Middle nasal concha
12. Ethmoid bone
13. Jugular process
14. Sphenoid bone
15. Mandible

Skull of pig – left orbit

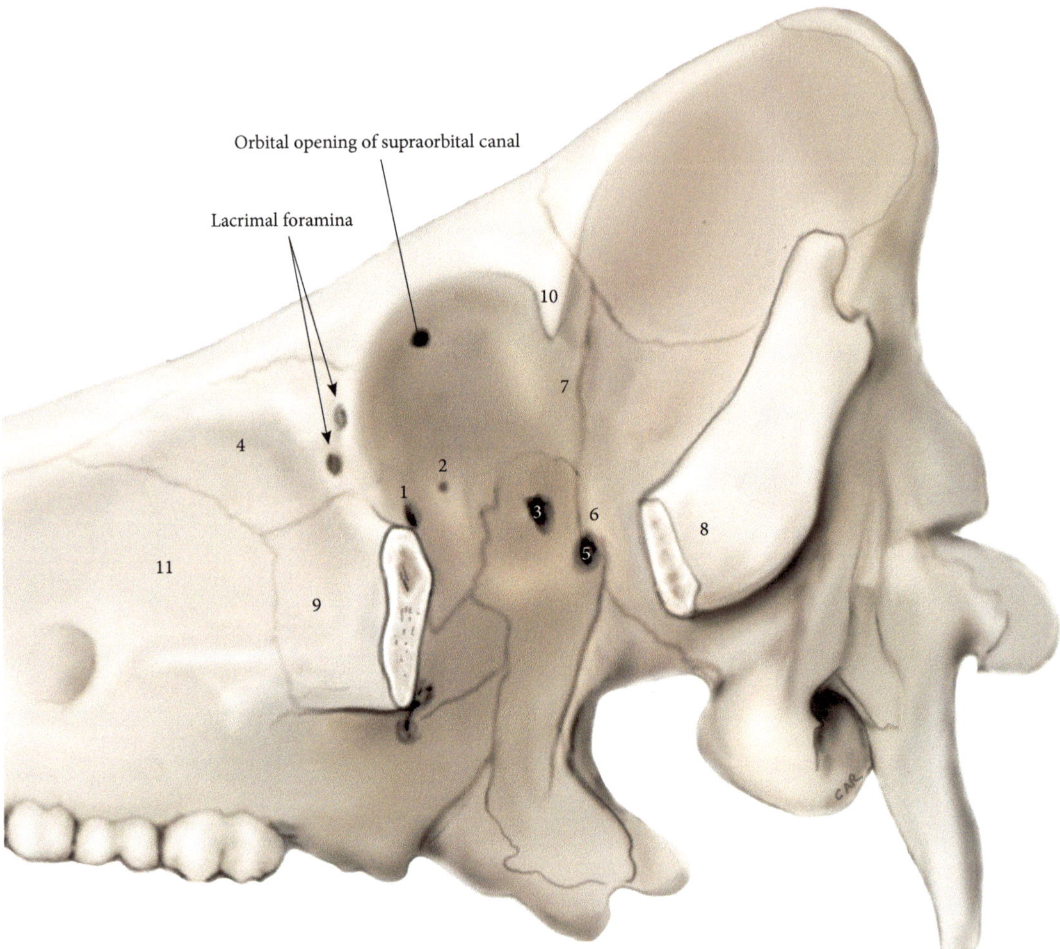

1. Fossa for ventral oblique muscle
2. Ethmoidal foramen
3. Optic canal
4. Lacrimal bone
5. Orbitorotundum foramen
6. Pterygoid crest
7. Orbitotemporal crest
8. Zygomatic process of temporal bone
9. Zygomatic bone
10. Zygomatic process of frontal bone
11. Maxilla

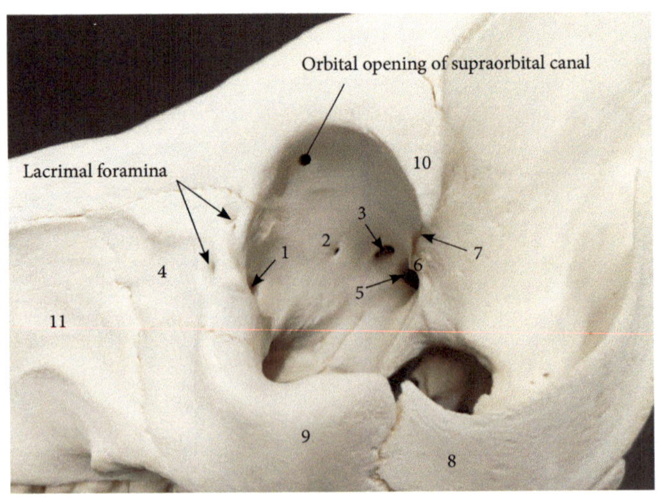

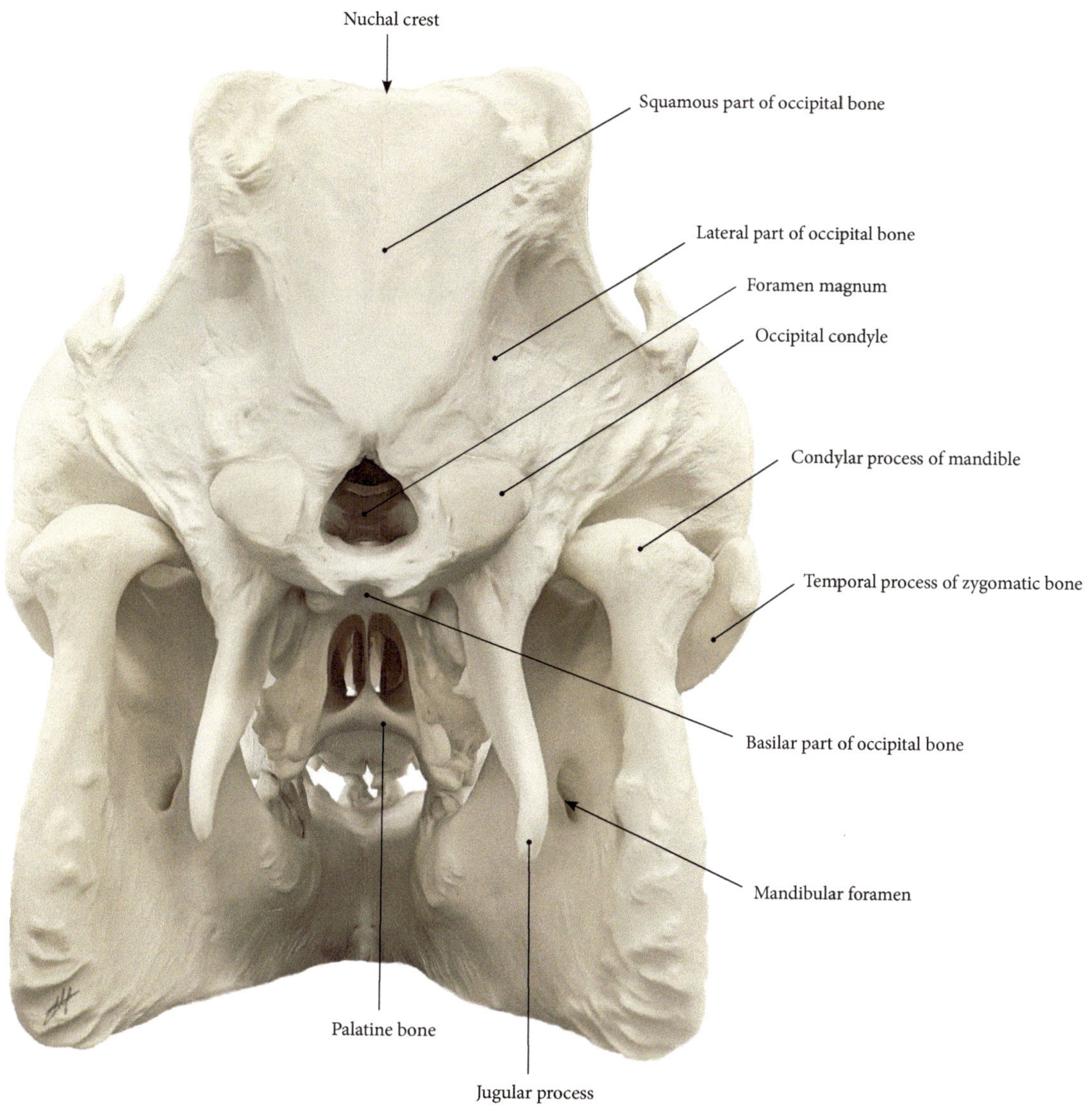
Nuchal crest
Squamous part of occipital bone
Lateral part of occipital bone
Foramen magnum
Occipital condyle
Condylar process of mandible
Temporal process of zygomatic bone
Basilar part of occipital bone
Mandibular foramen
Palatine bone
Jugular process

Paranasal sinuses

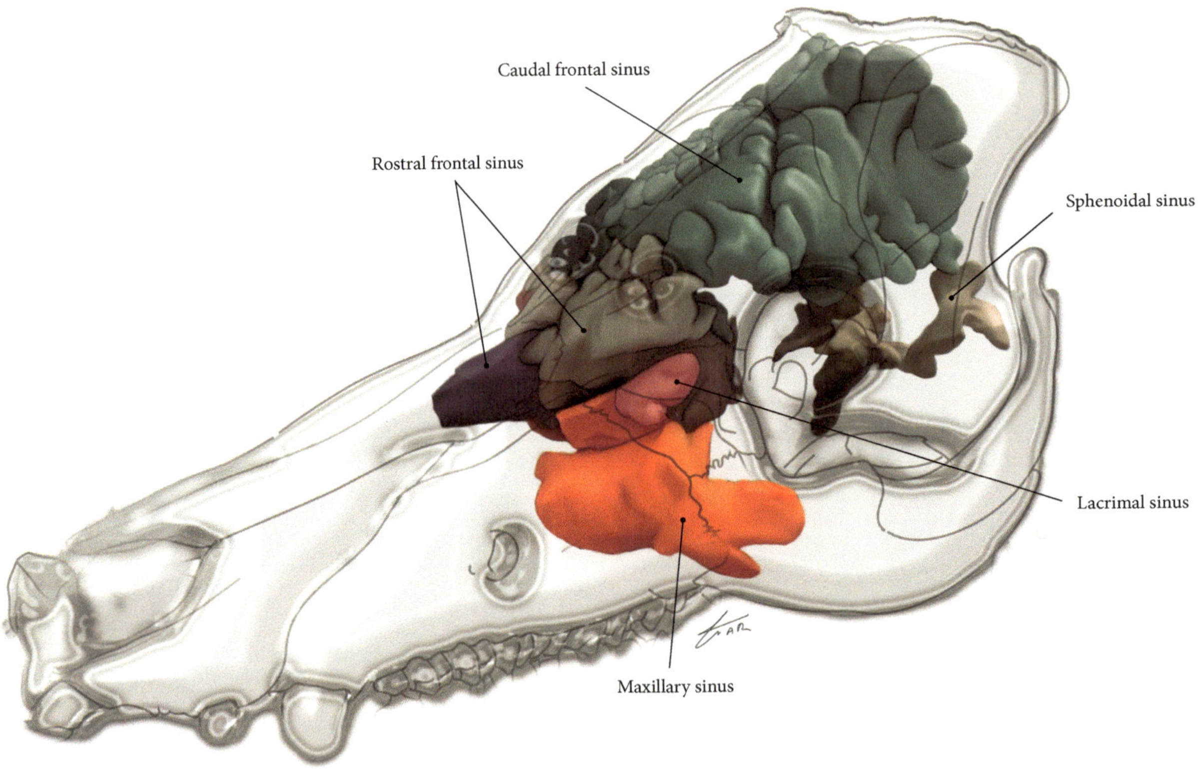

Hyoid bone of pig – rostrolateral

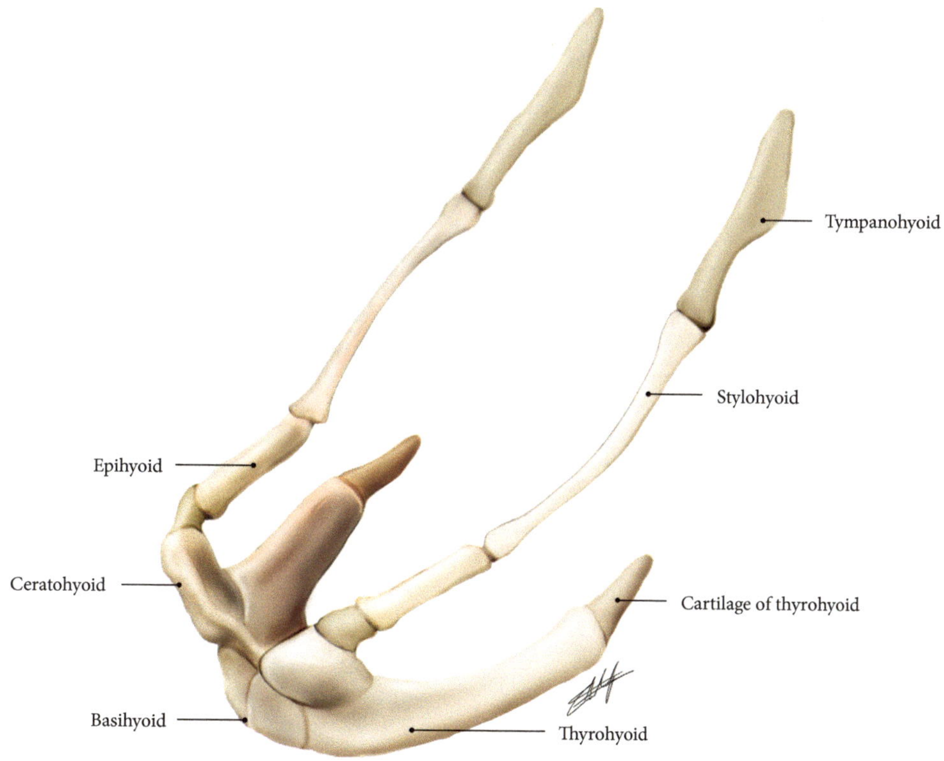

Atlas, fourth cervical vertebra, and axis

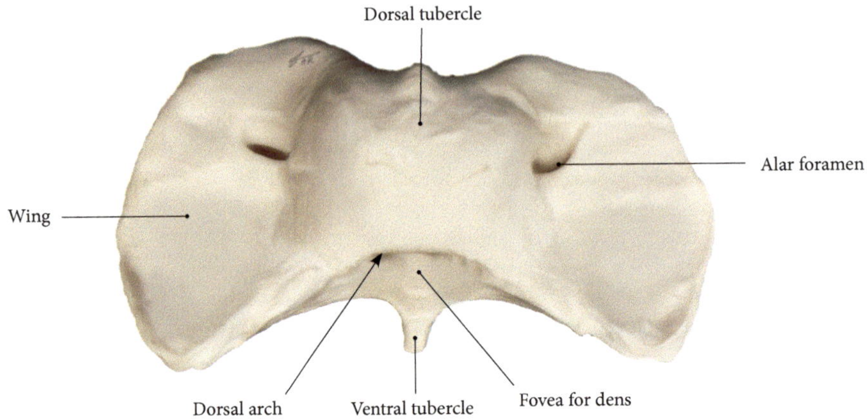

Atlas – caudodorsal view

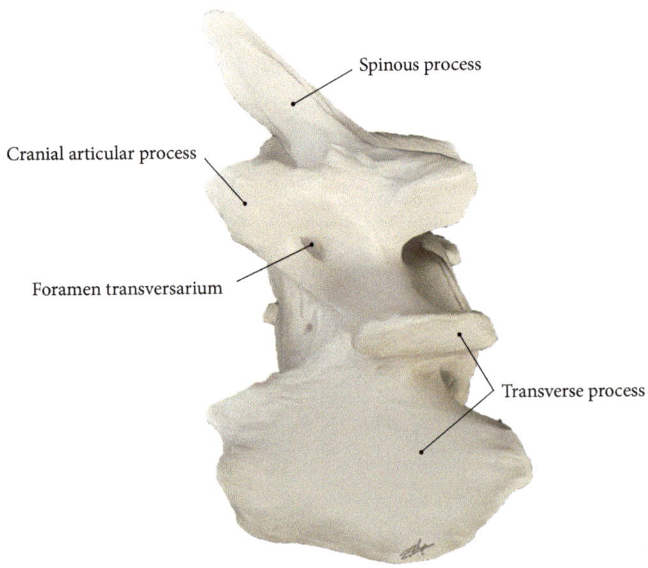

Fourth cervical vertebra – left view

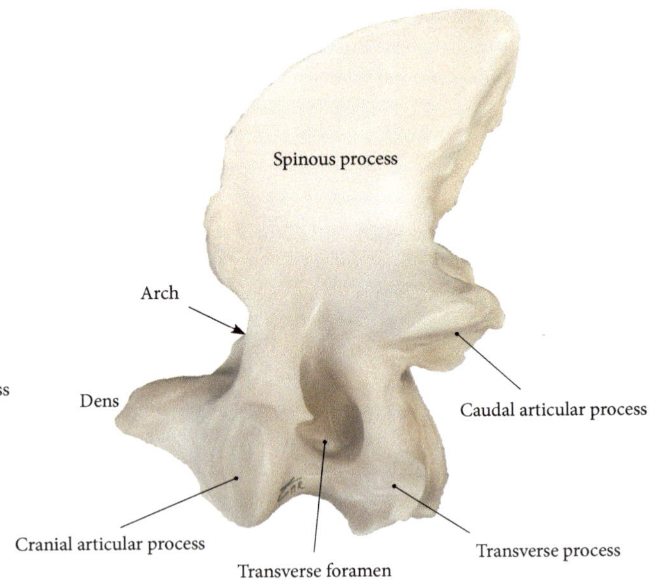

Axis – left view

Thoracic vertebrae – lateral view

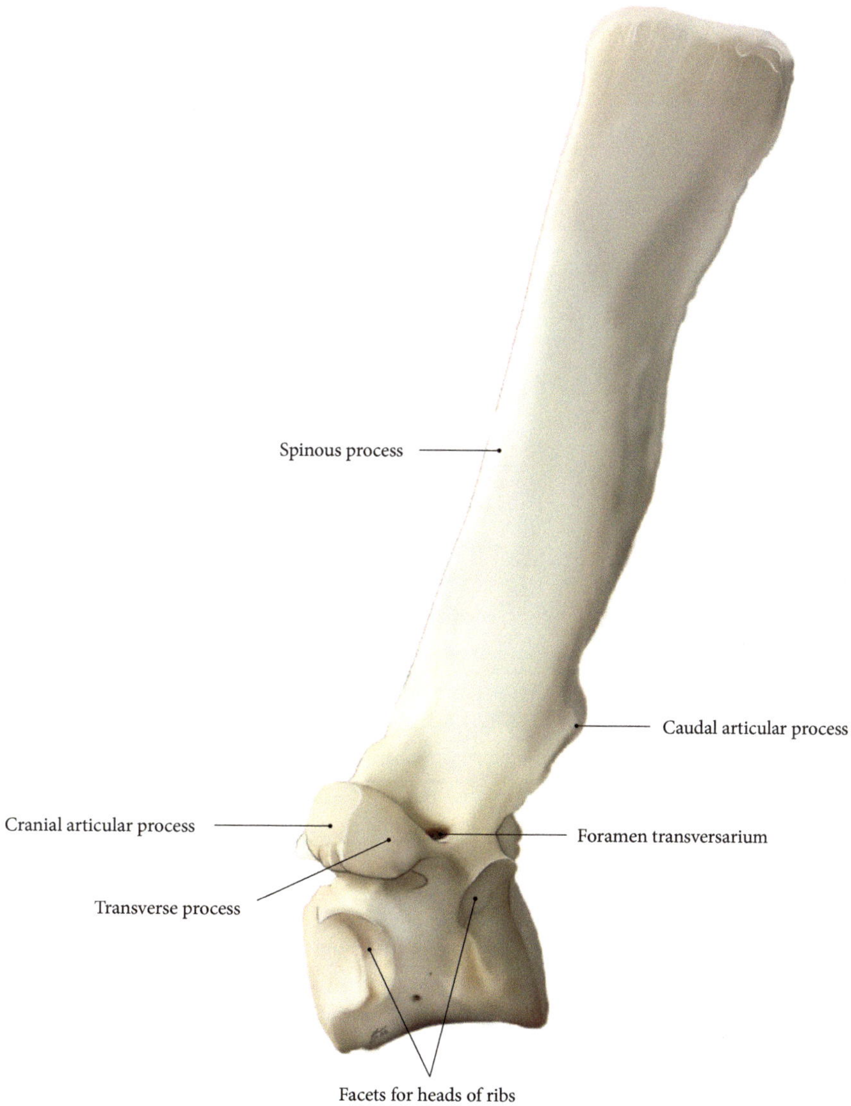

Lumbar vertebrae

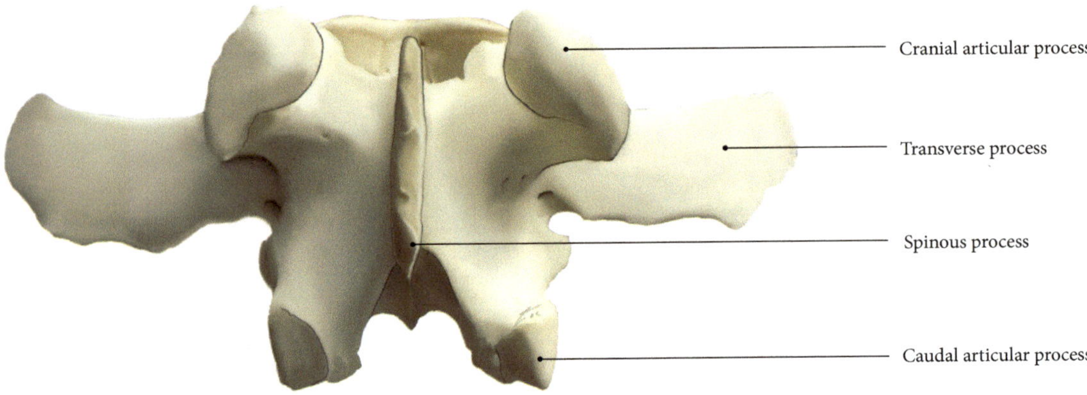

Dorsal view

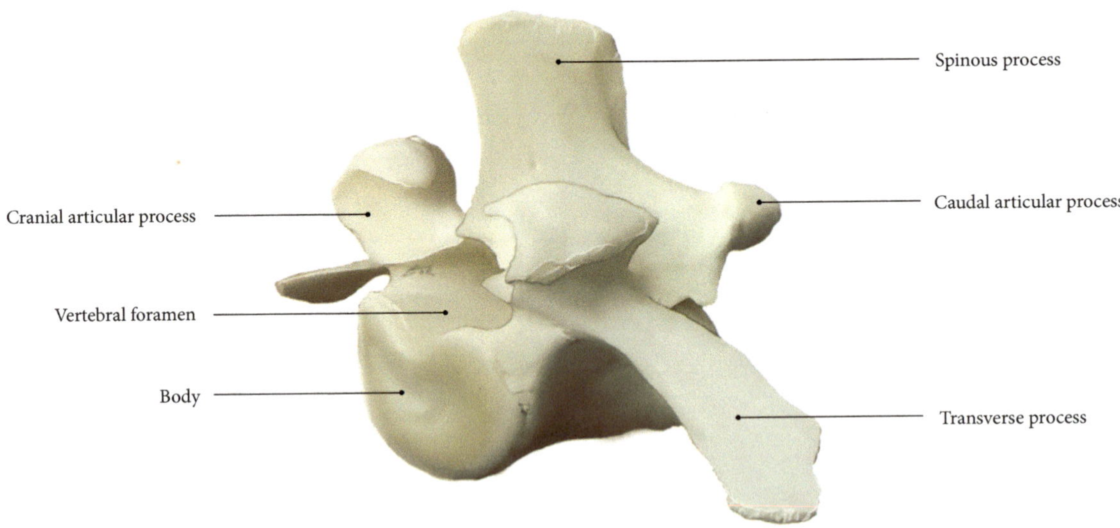

Caudolateral view

Sacrum and first caudal vertebra of pig

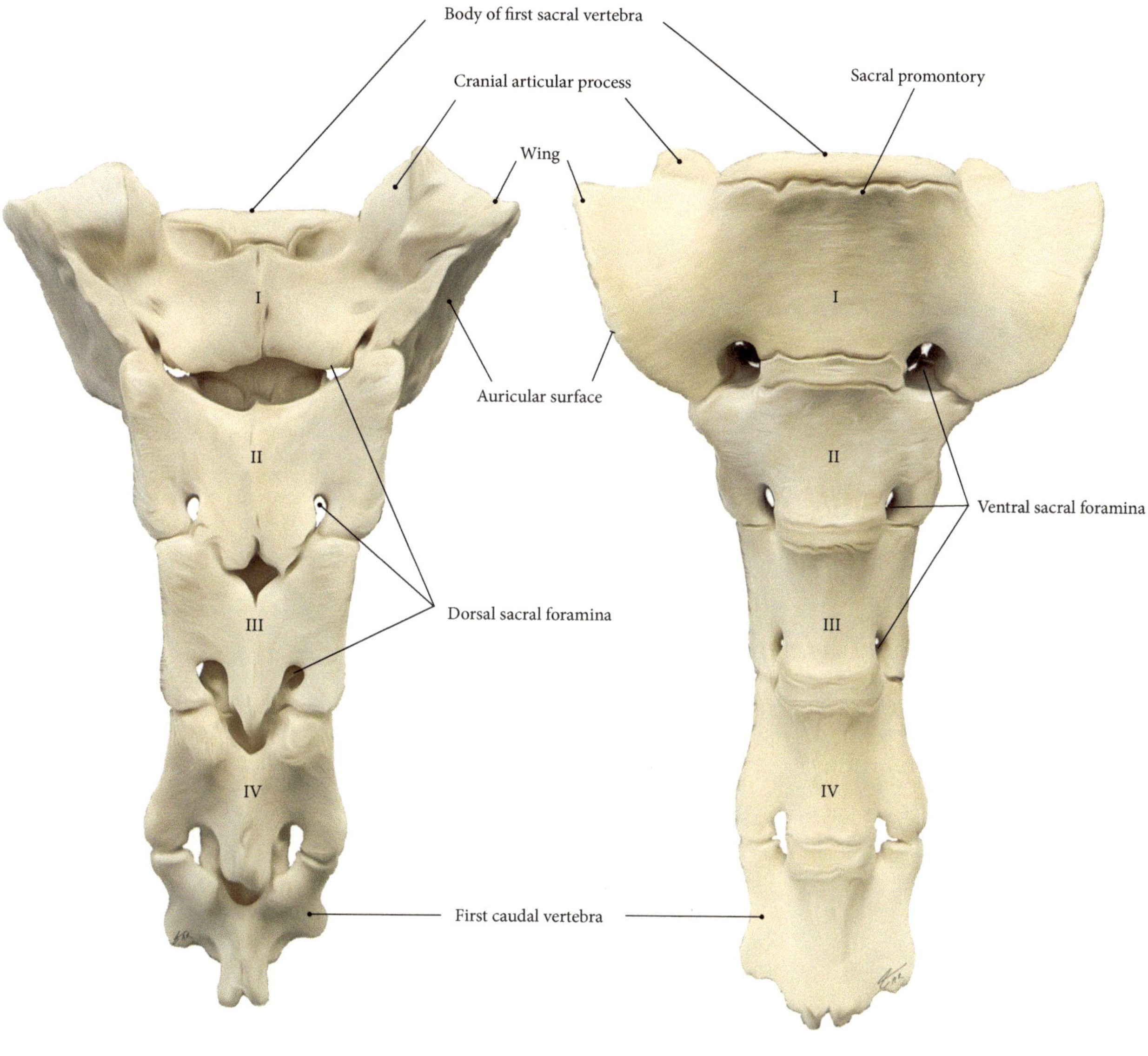

Dorsal view

Ventral view

Rib of pig

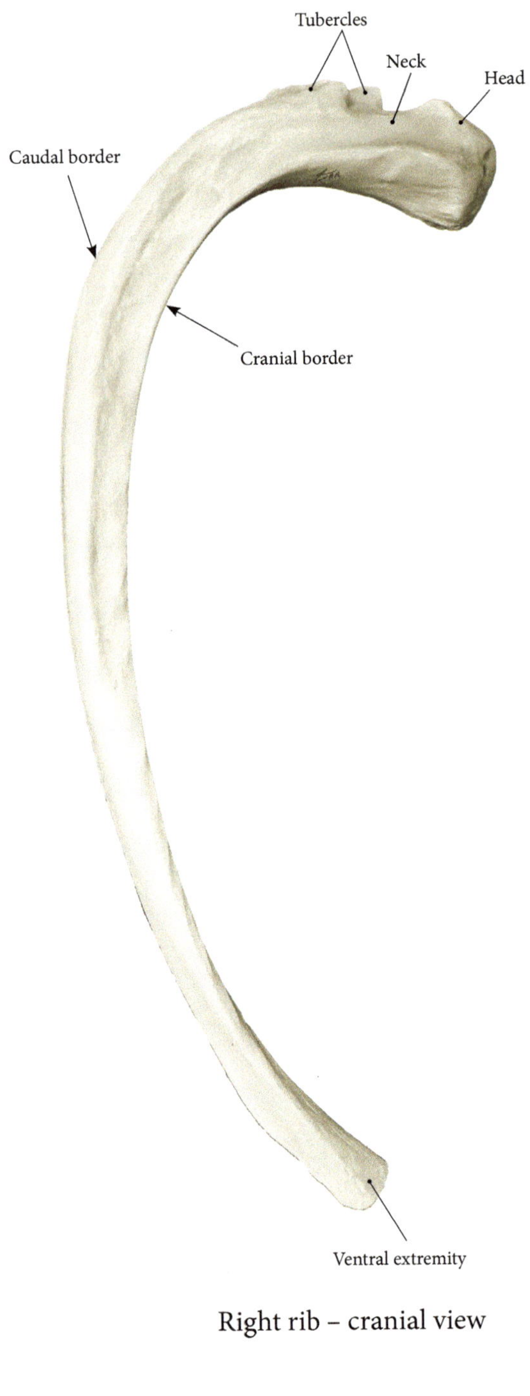

Right rib – cranial view

Right rib – caudal view

Sternum – dorsal view

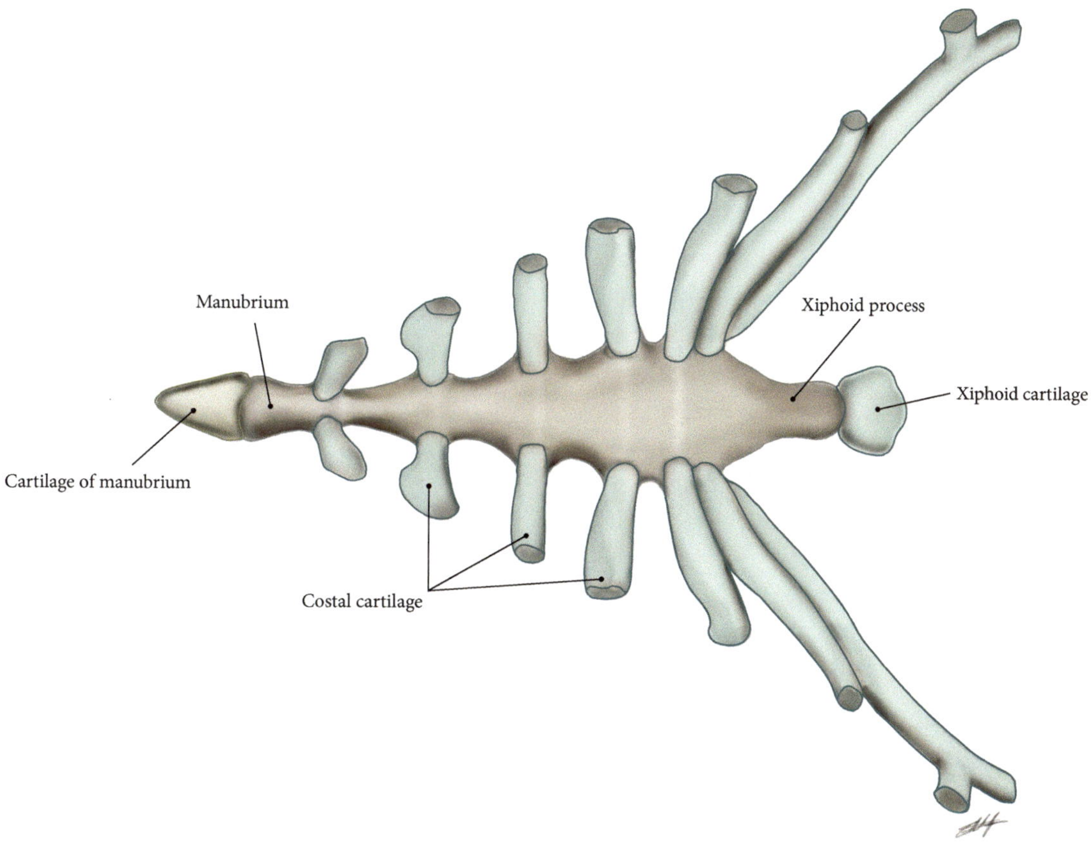

Sternum – lateral view

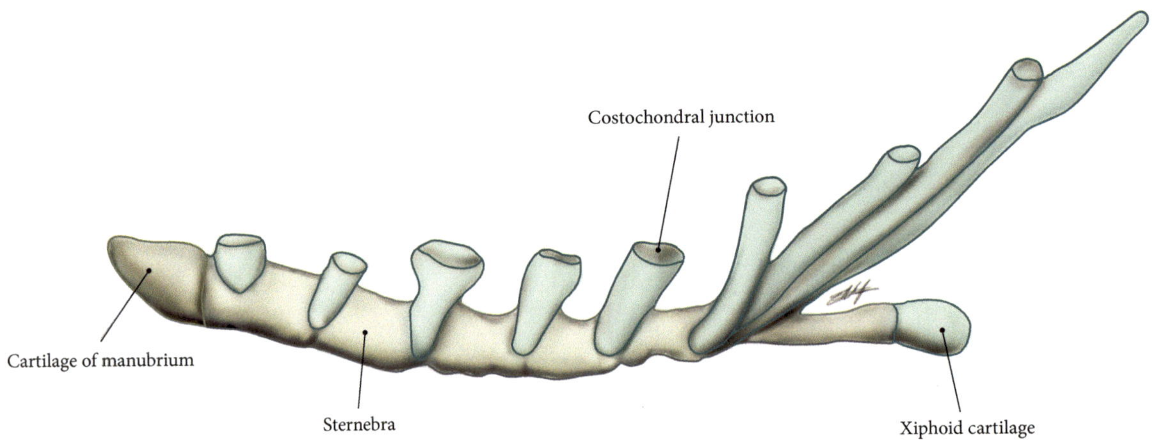

Bones of the right thoracic limb

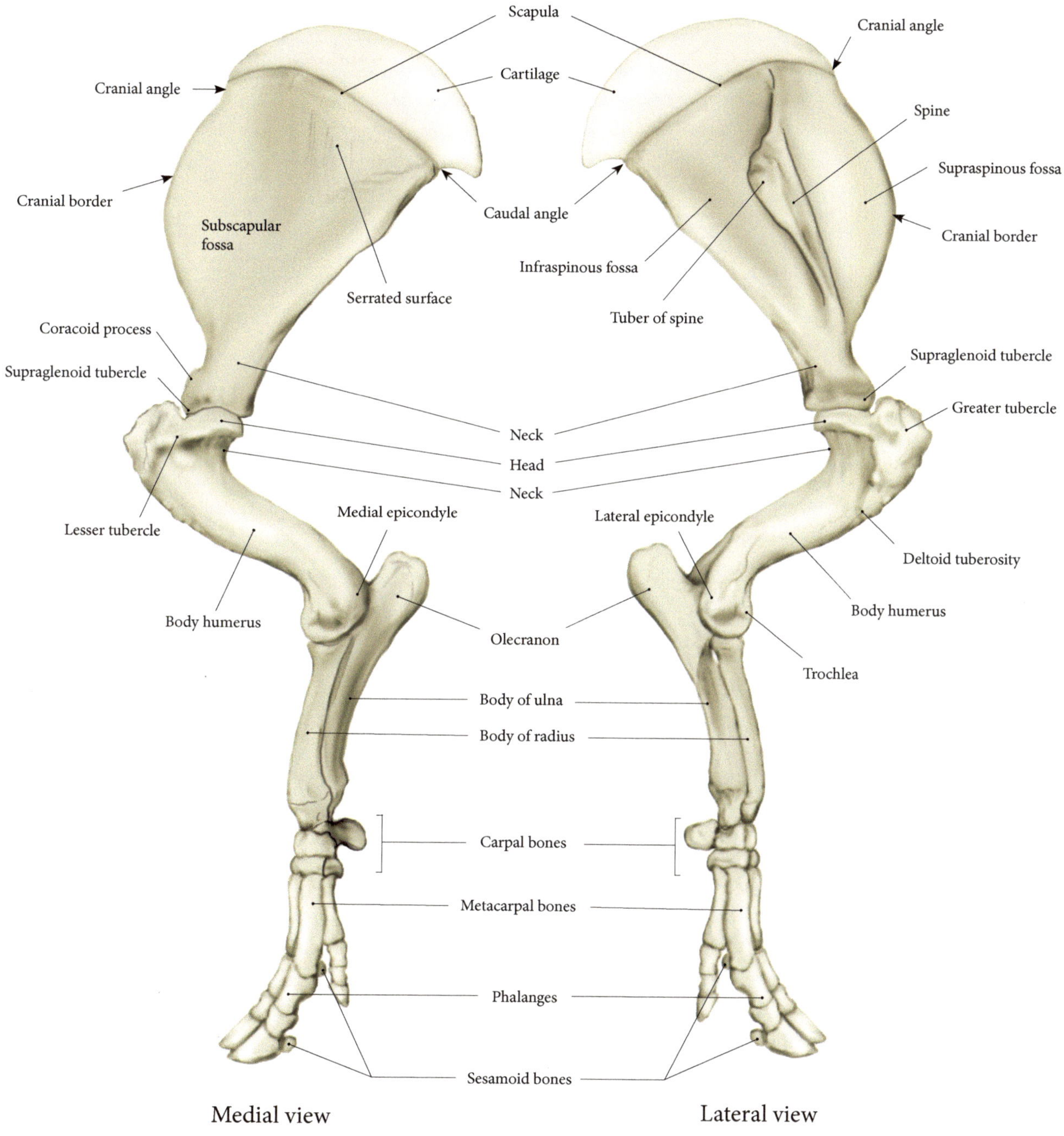

Bones of the right pelvic limb

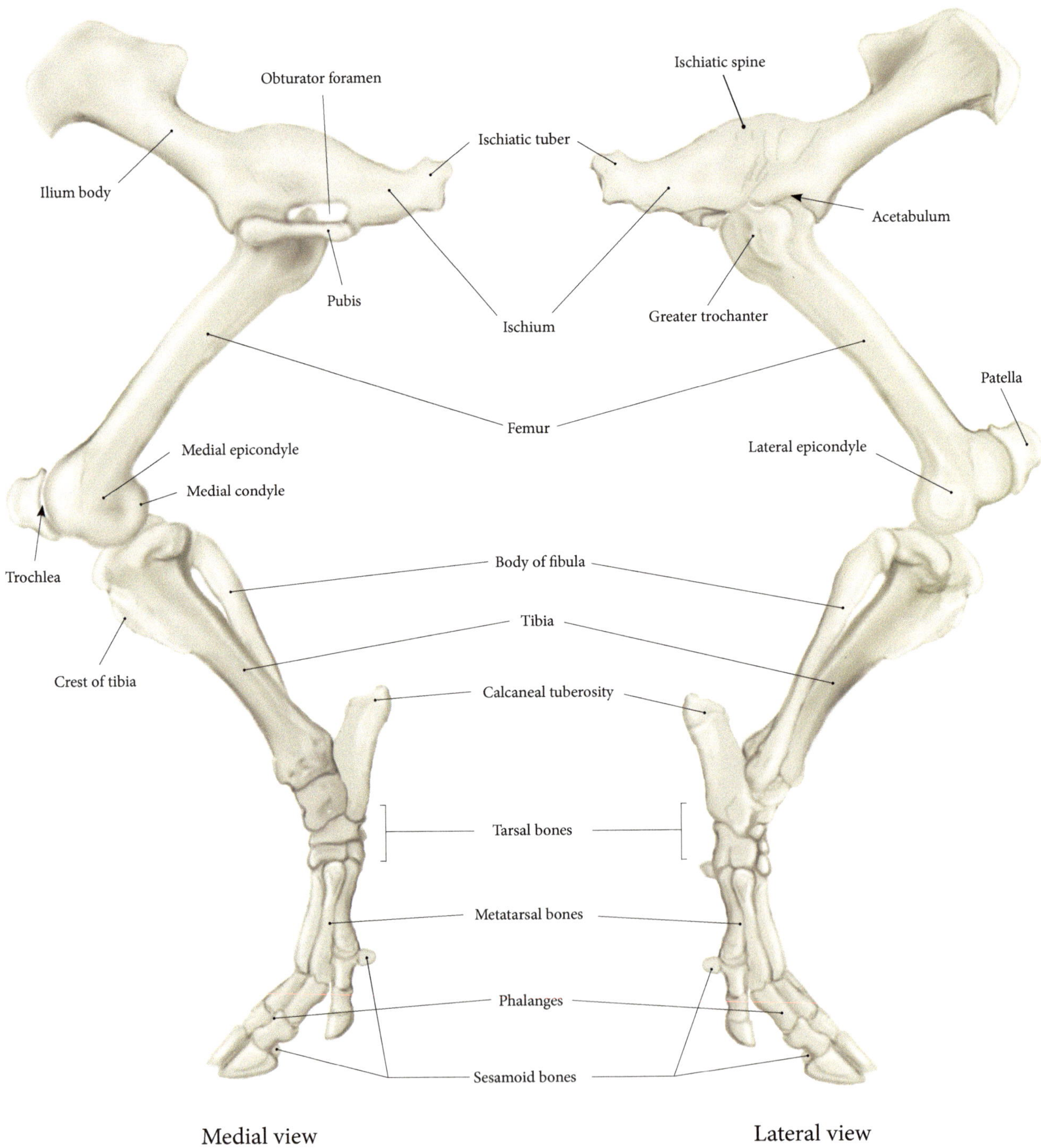

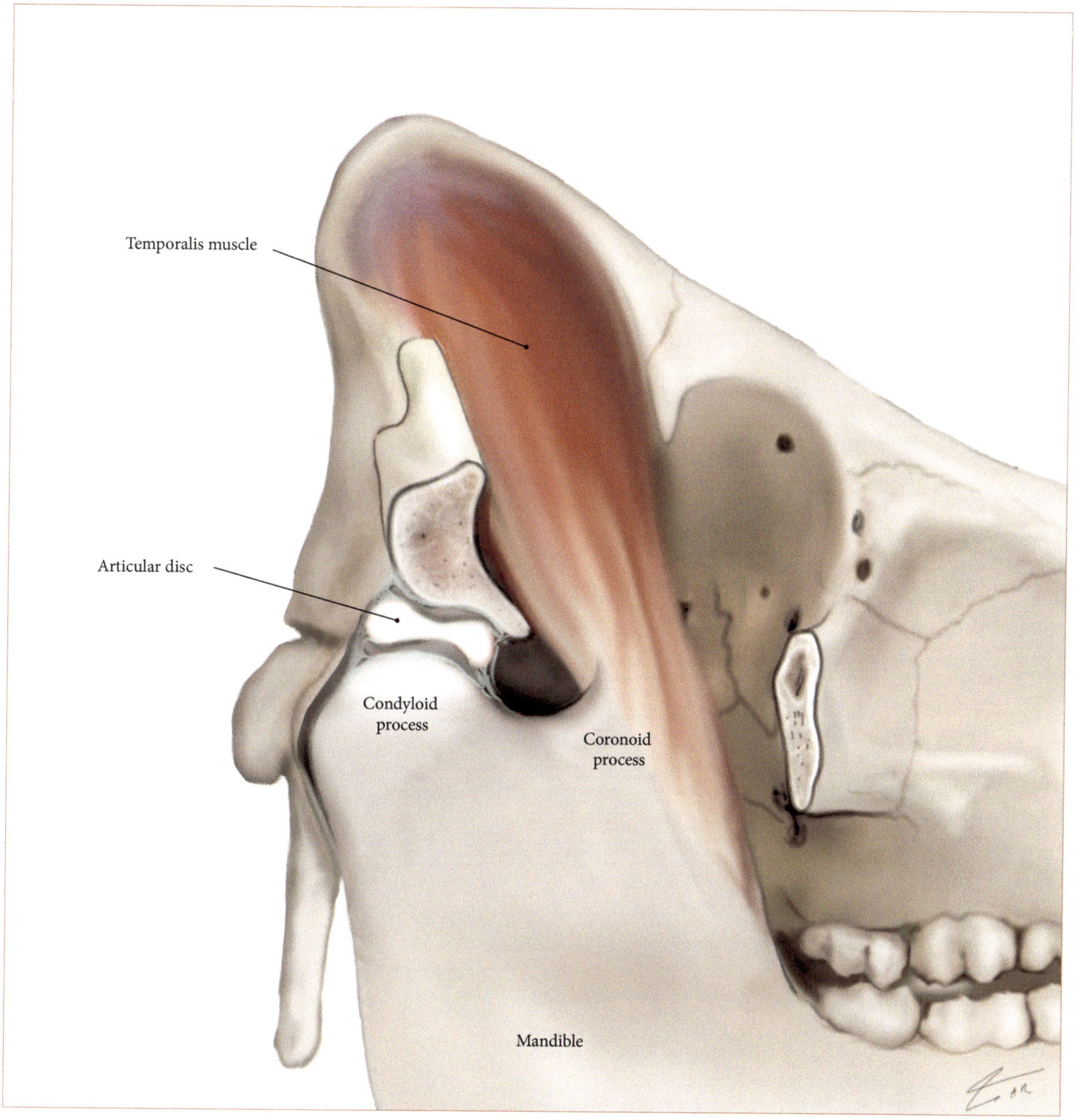
Temporalis muscle
Articular disc
Condyloid process
Coronoid process
Mandible

Glenohumeral joint – cranial view

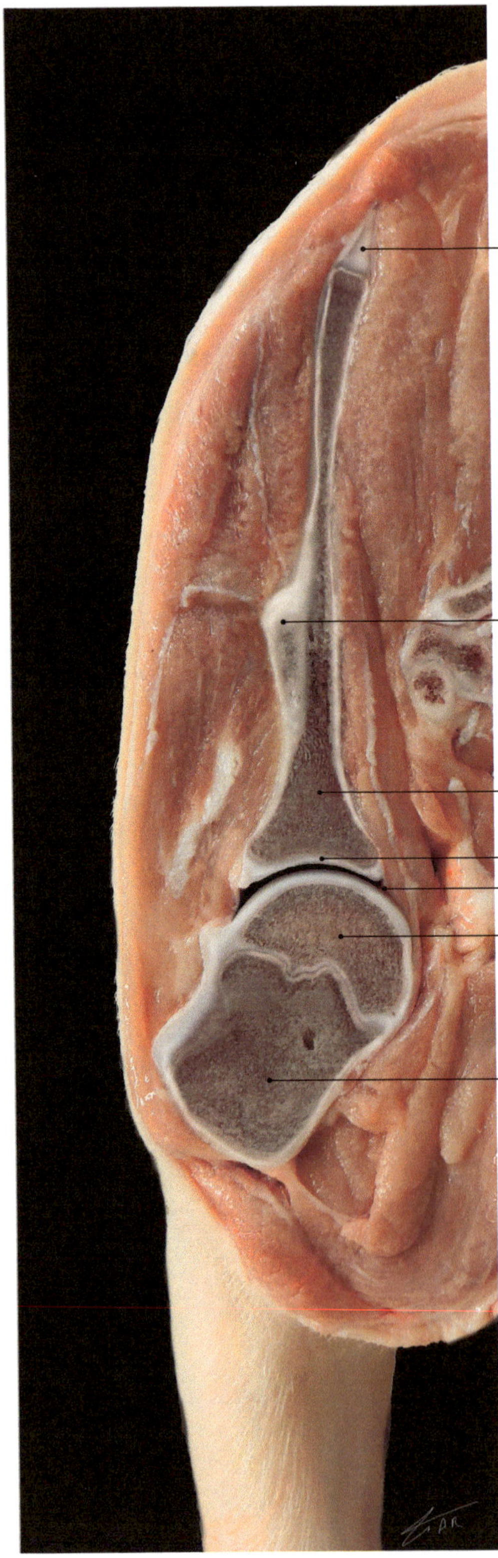

Left stifle joint – medial aspect

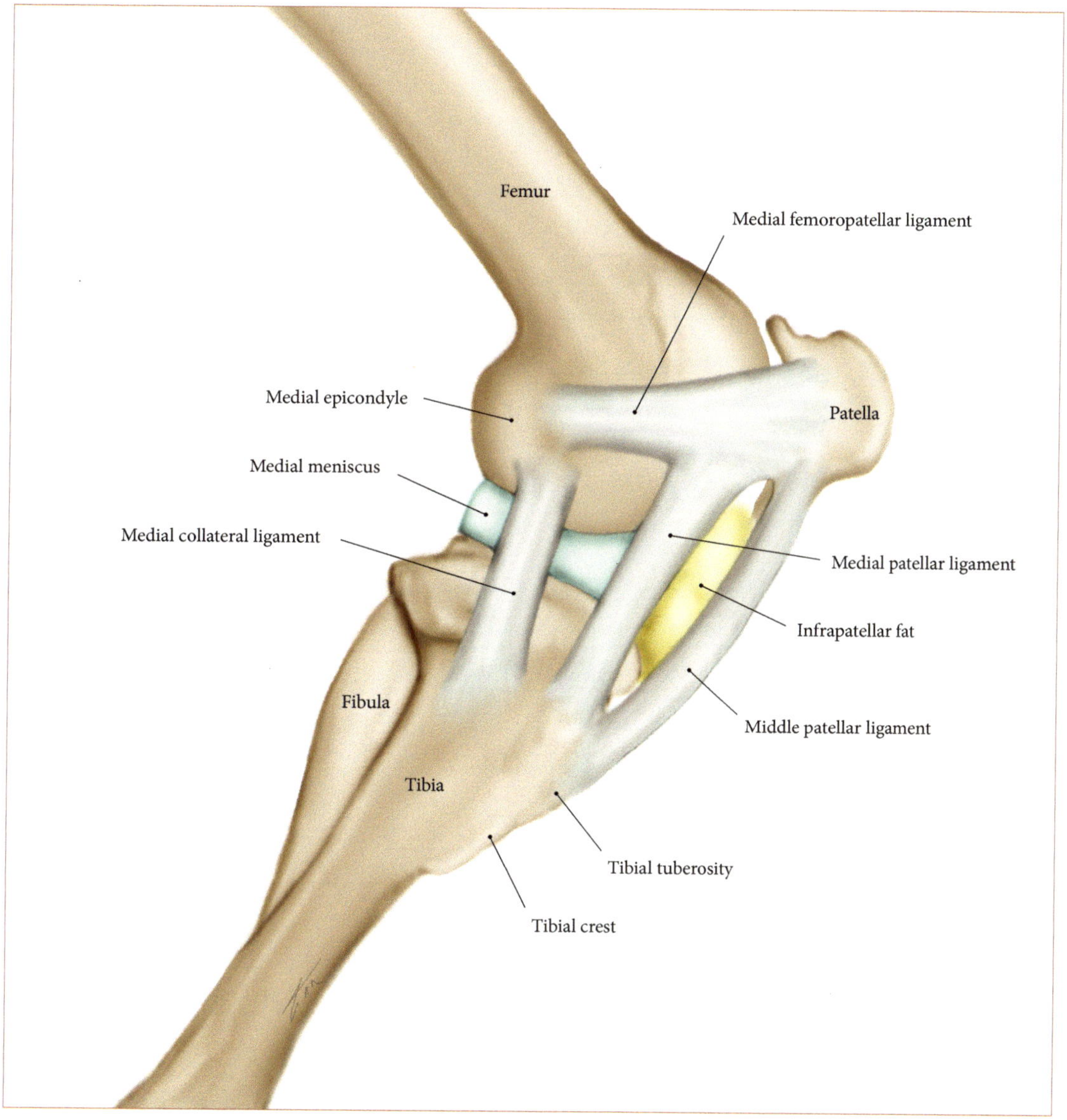

Left stifle joint – lateral aspect

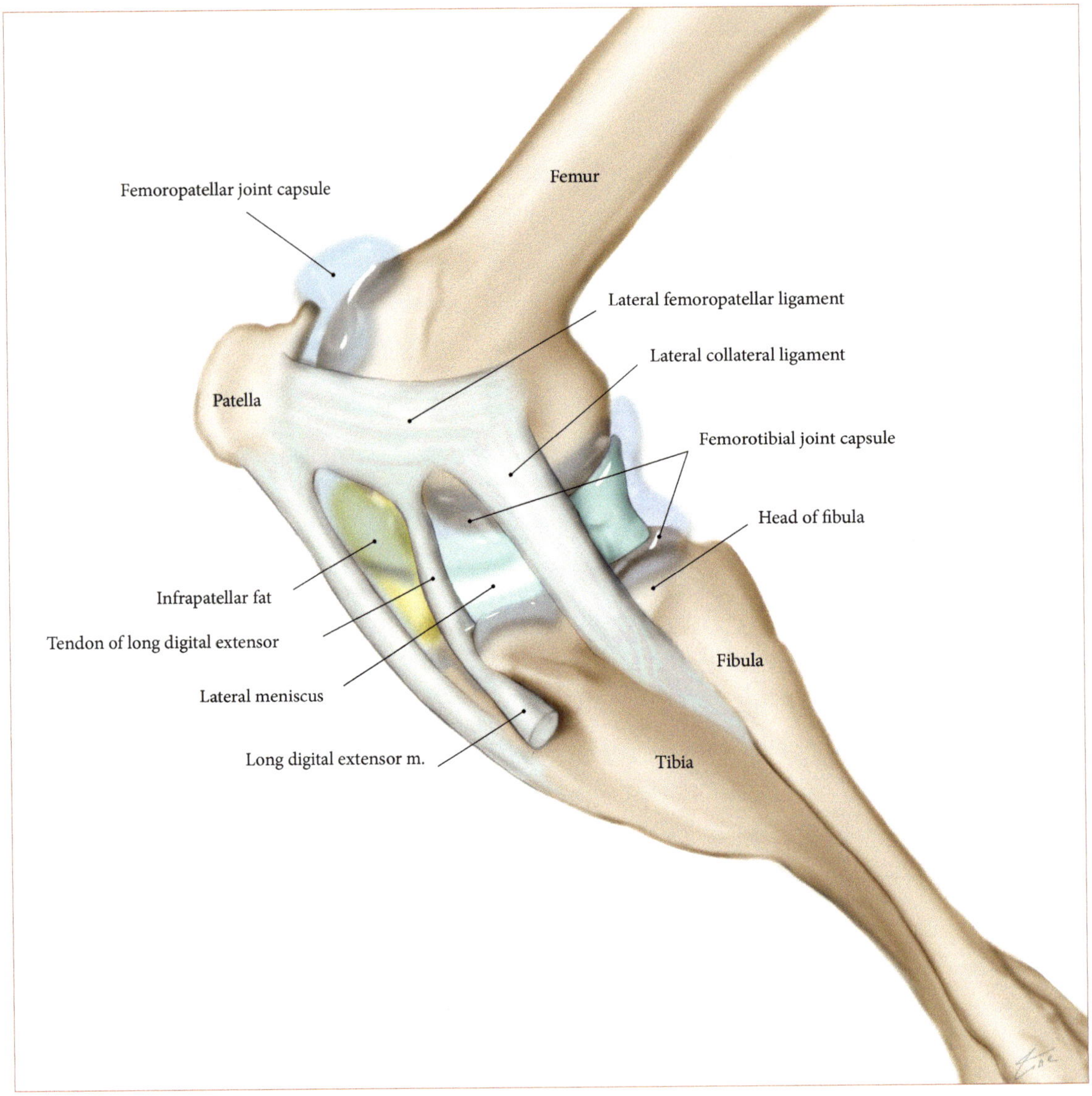

Right stifle joint – caudal view

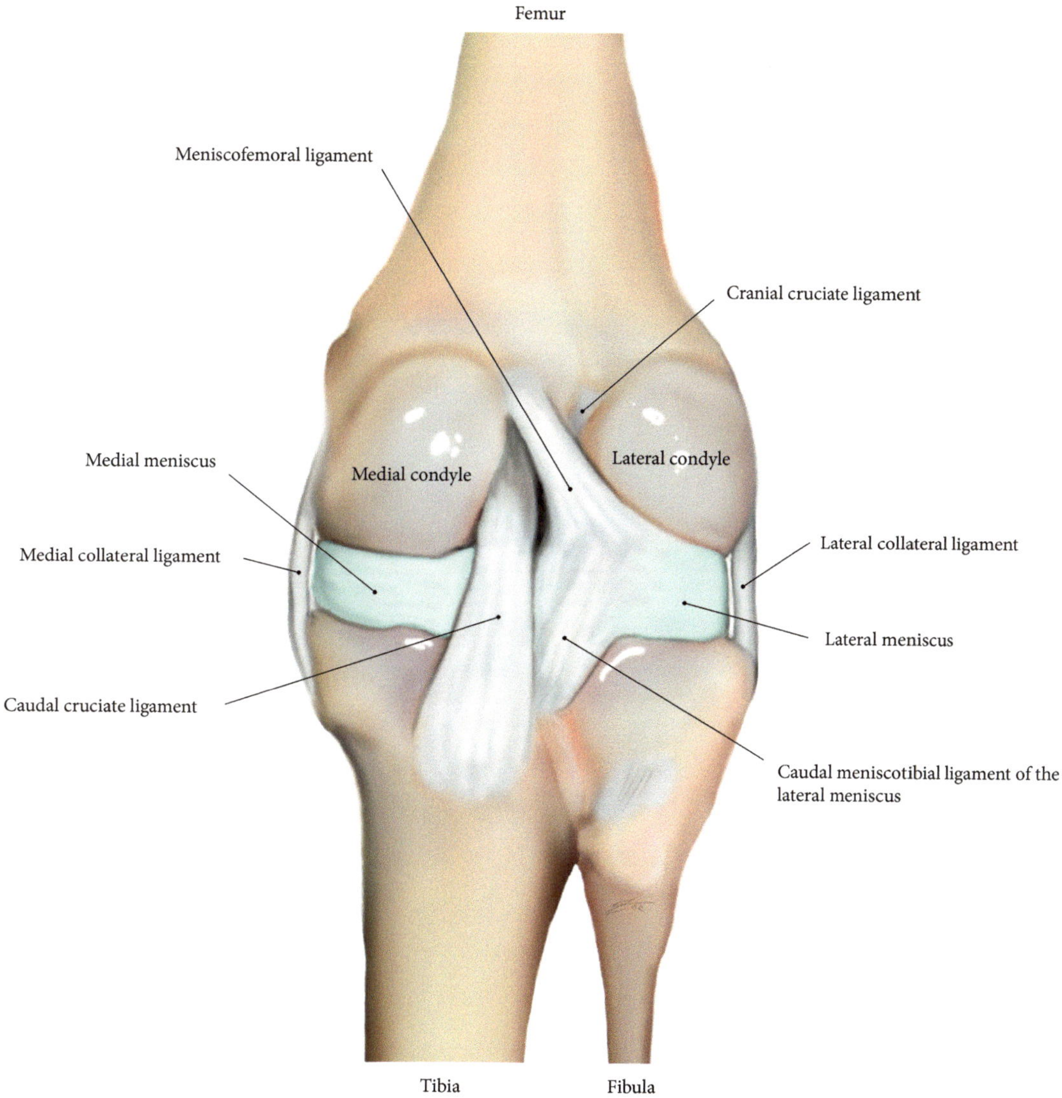

Right stifle joint – cranial flexed view

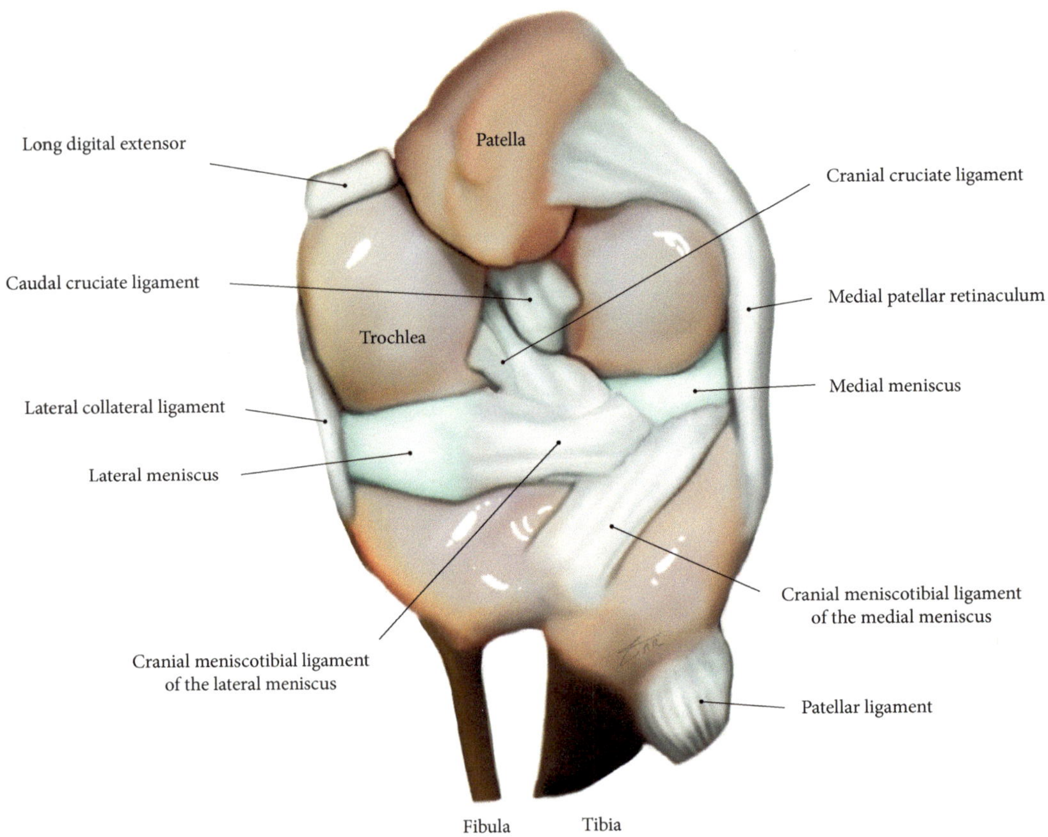

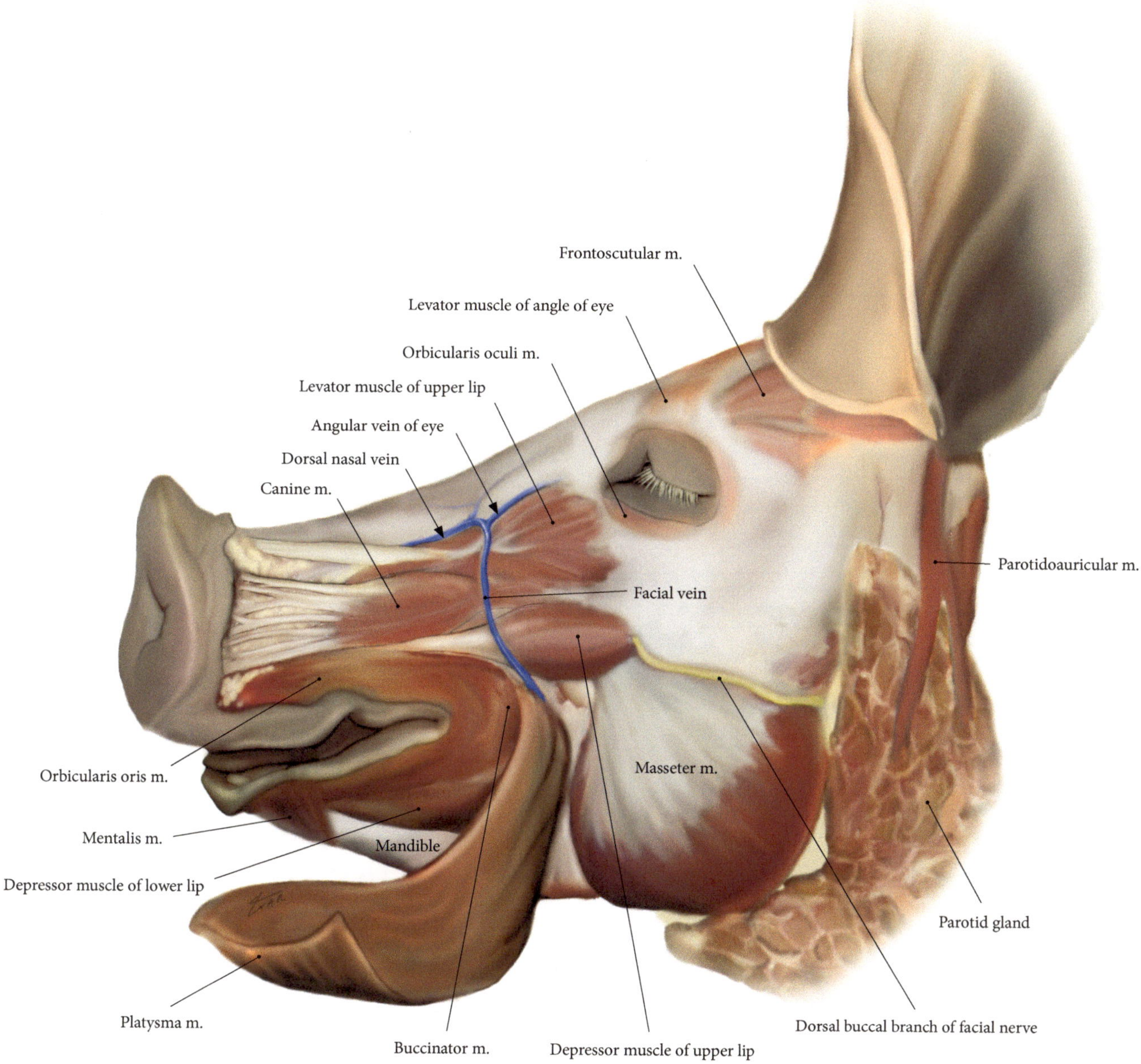
Frontoscutular m.
Levator muscle of angle of eye
Orbicularis oculi m.
Levator muscle of upper lip
Angular vein of eye
Dorsal nasal vein
Canine m.
Parotidoauricular m.
Facial vein
Orbicularis oris m.
Masseter m.
Mentalis m.
Mandible
Depressor muscle of lower lip
Parotid gland
Platysma m.
Dorsal buccal branch of facial nerve
Buccinator m.
Depressor muscle of upper lip

Mastication muscles

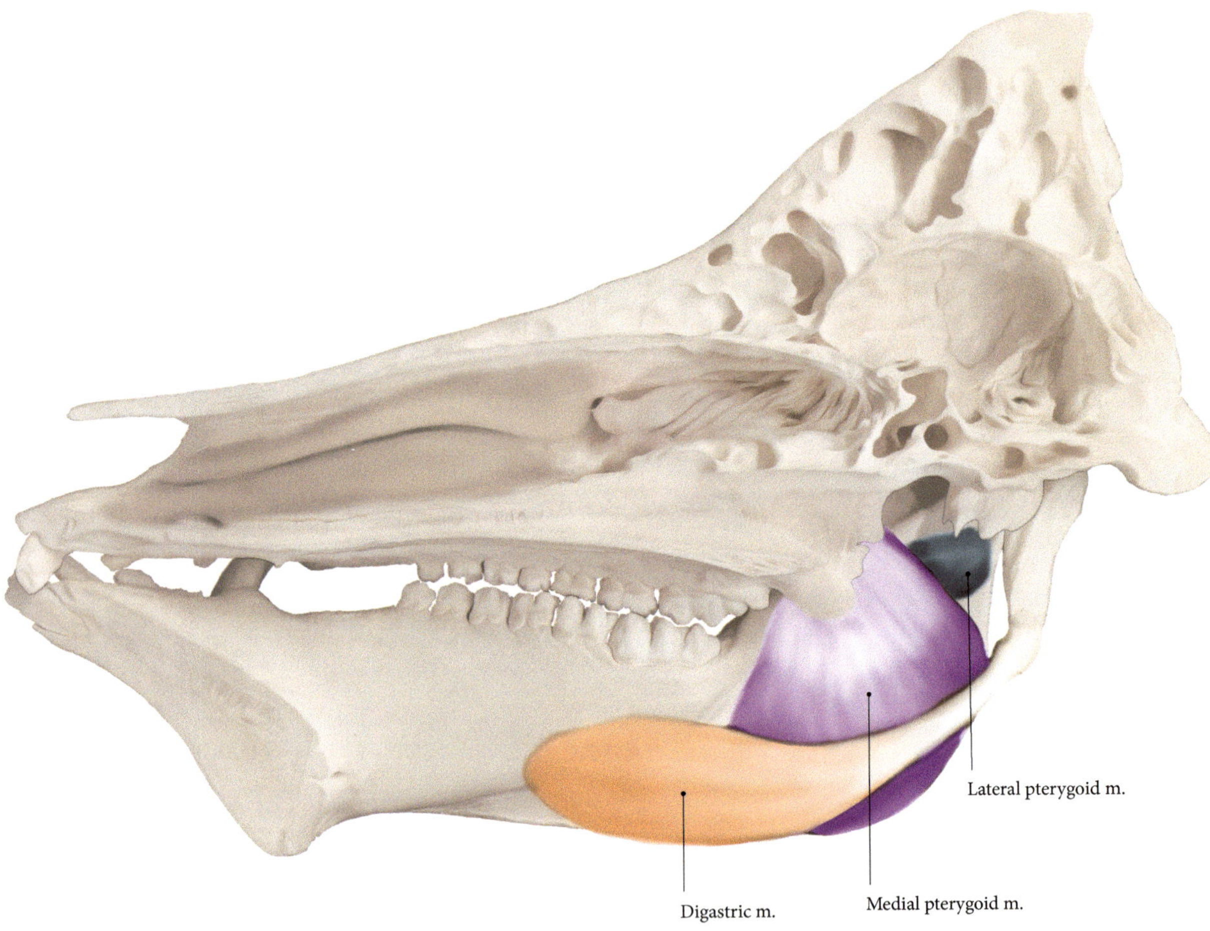

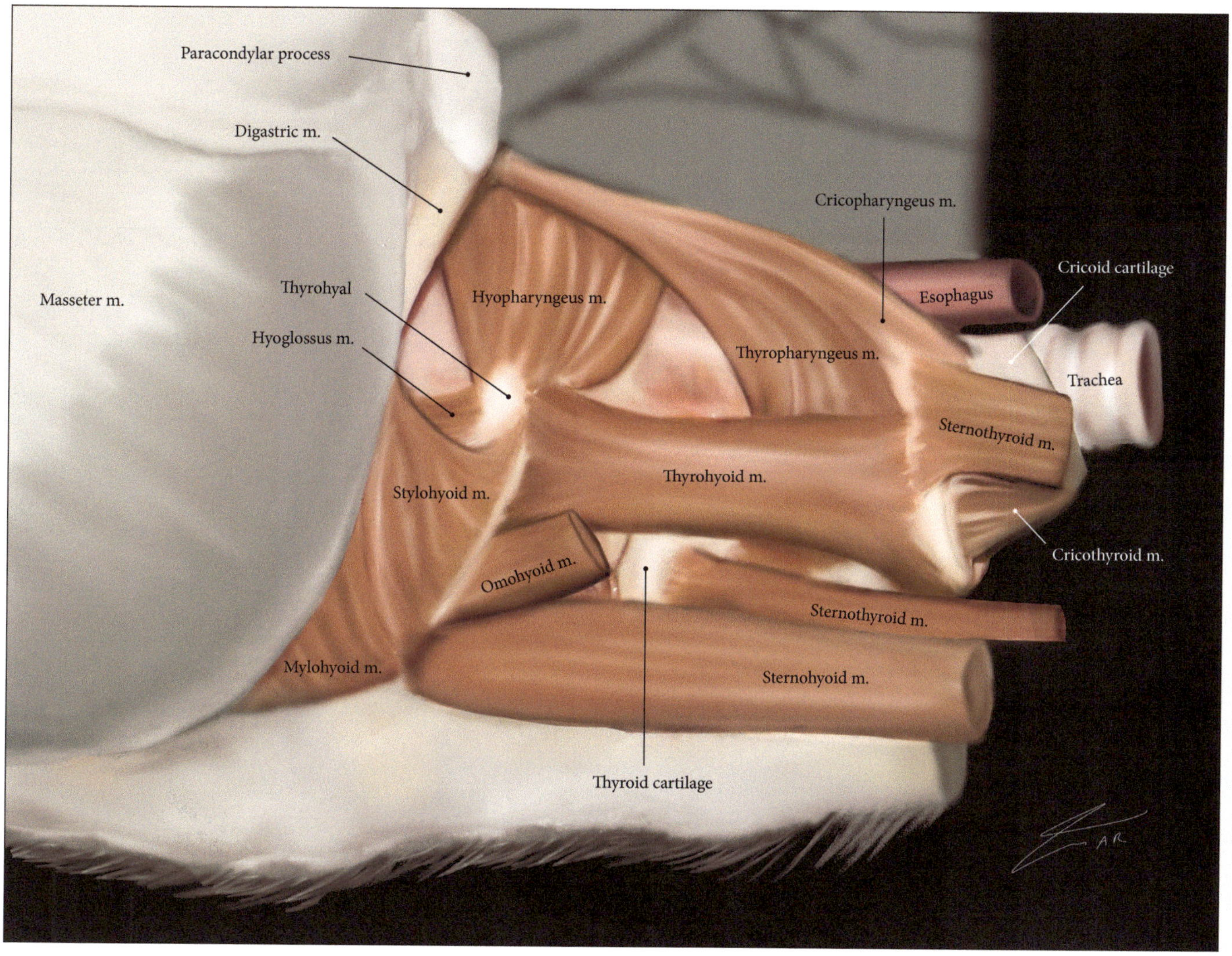
Paracondylar process
Digastric m.
Cricopharyngeus m.
Cricoid cartilage
Masseter m.
Thyrohyal
Hyopharyngeus m.
Esophagus
Hyoglossus m.
Thyropharyngeus m.
Trachea
Sternothyroid m.
Thyrohyoid m.
Stylohyoid m.
Cricothyroid m.
Omohyoid m.
Sternothyroid m.
Mylohyoid m.
Sternohyoid m.
Thyroid cartilage

Laryngeal muscles – ventral aspect, superficial and deep dissections

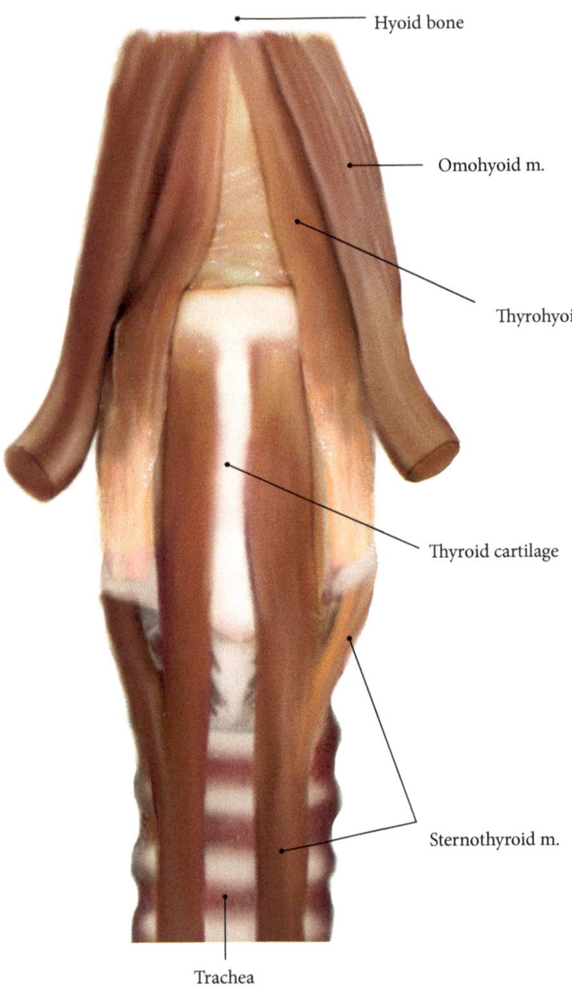

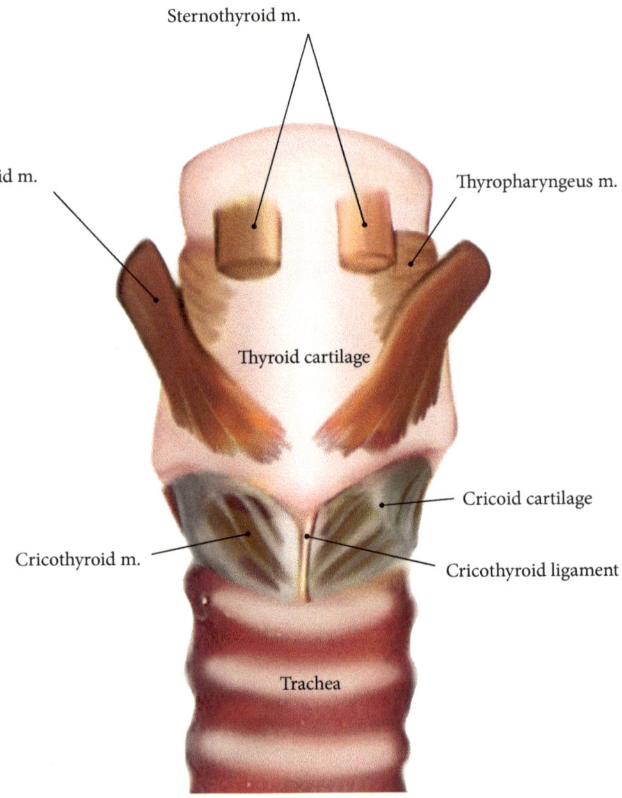

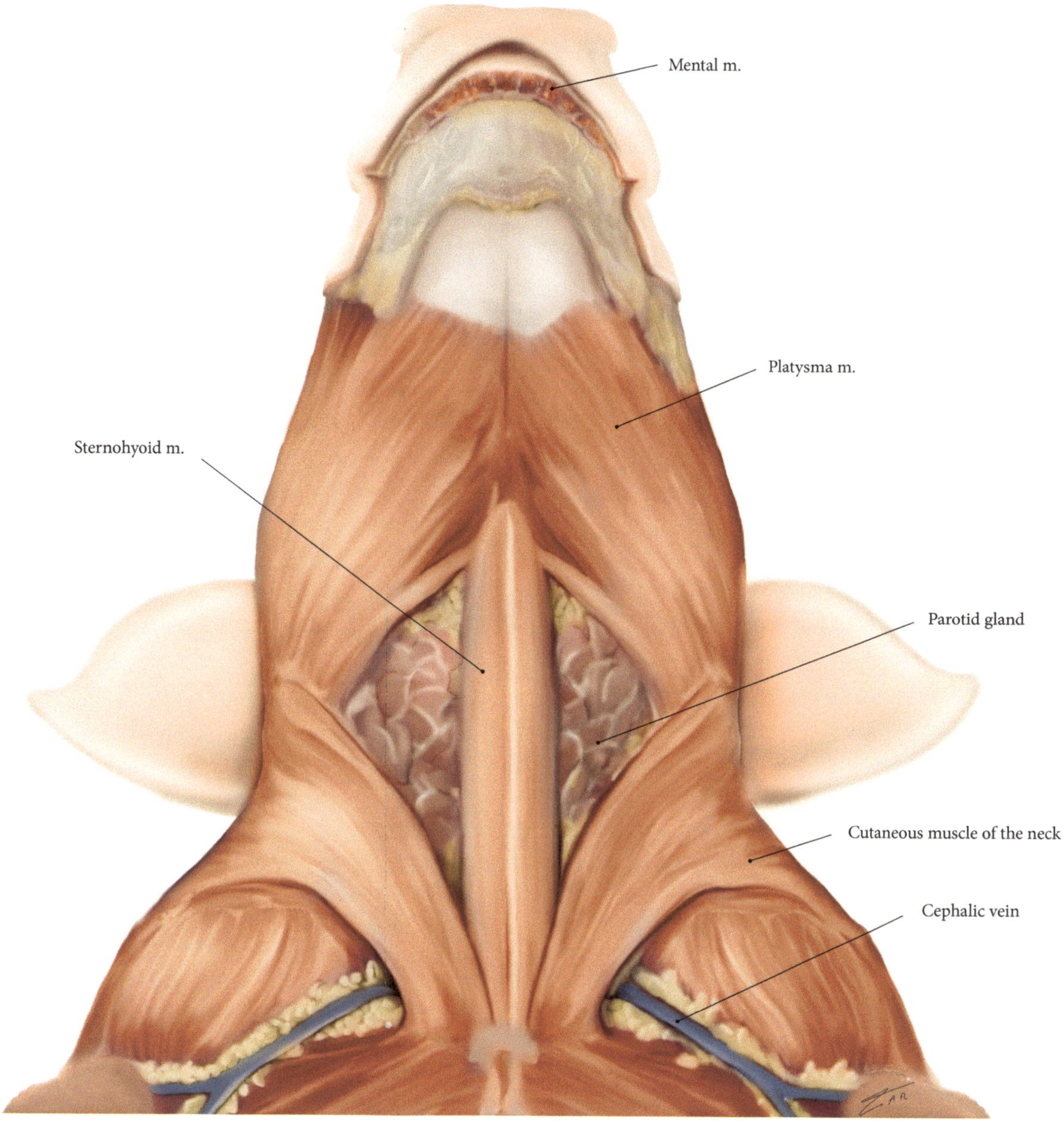
Mental m.
Platysma m.
Sternohyoid m.
Parotid gland
Cutaneous muscle of the neck
Cephalic vein

Dissection of the neck – ventral aspect

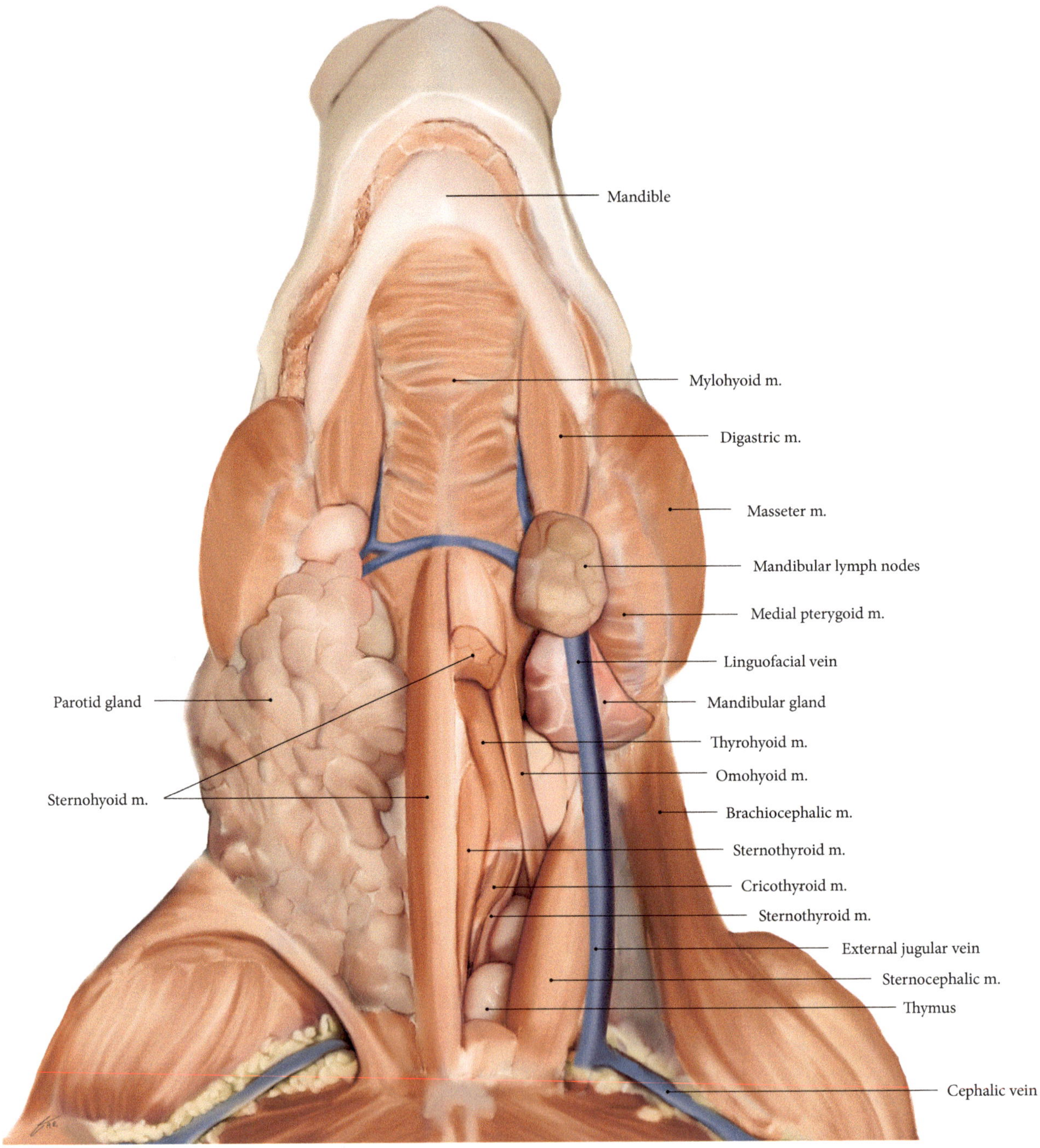

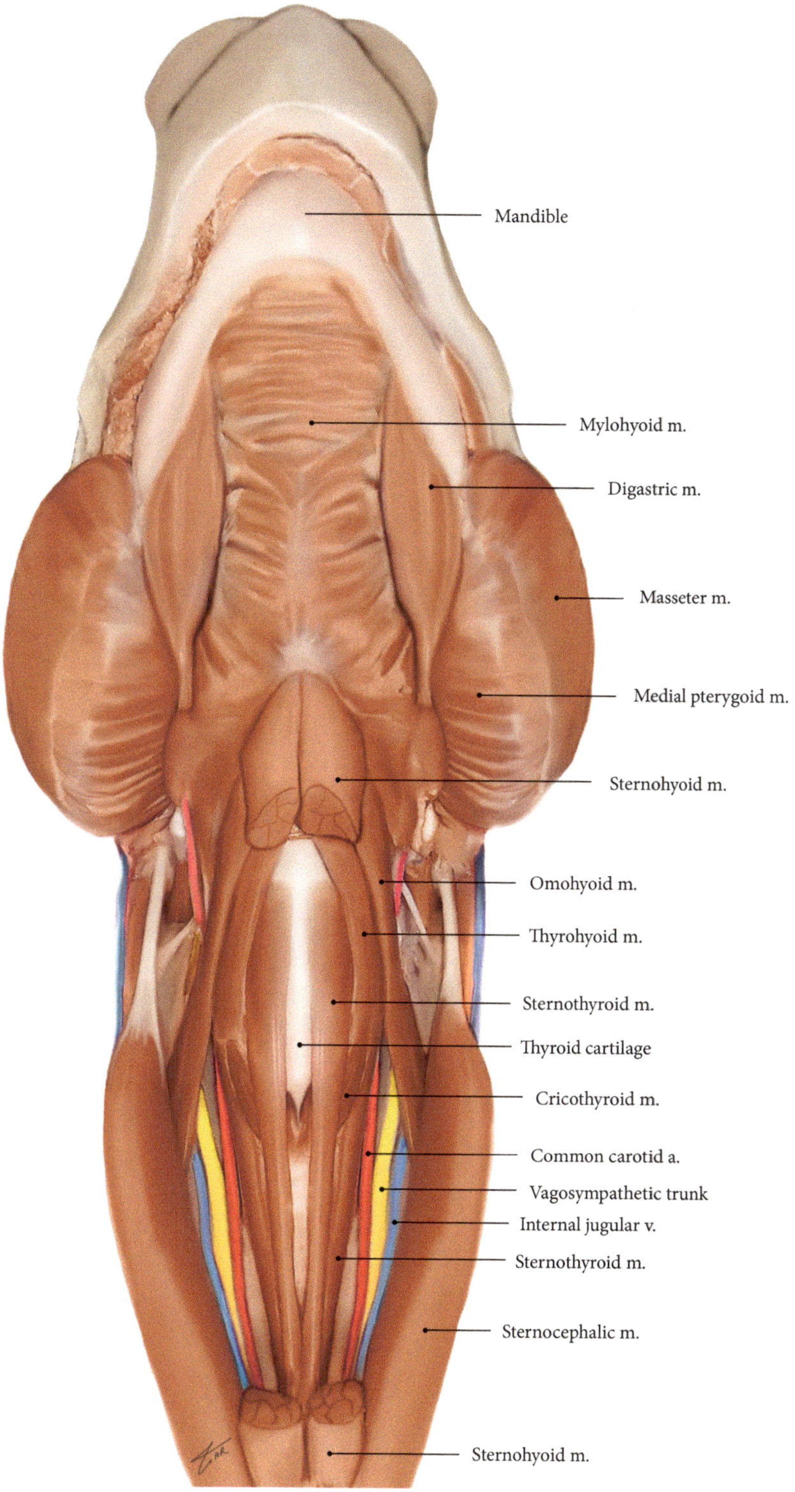
Mandible
Mylohyoid m.
Digastric m.
Masseter m.
Medial pterygoid m.
Sternohyoid m.
Omohyoid m.
Thyrohyoid m.
Sternothyroid m.
Thyroid cartilage
Cricothyroid m.
Common carotid a.
Vagosympathetic trunk
Internal jugular v.
Sternothyroid m.
Sternocephalic m.
Sternohyoid m.

Muscles of the neck – superficial dissection, lateral view

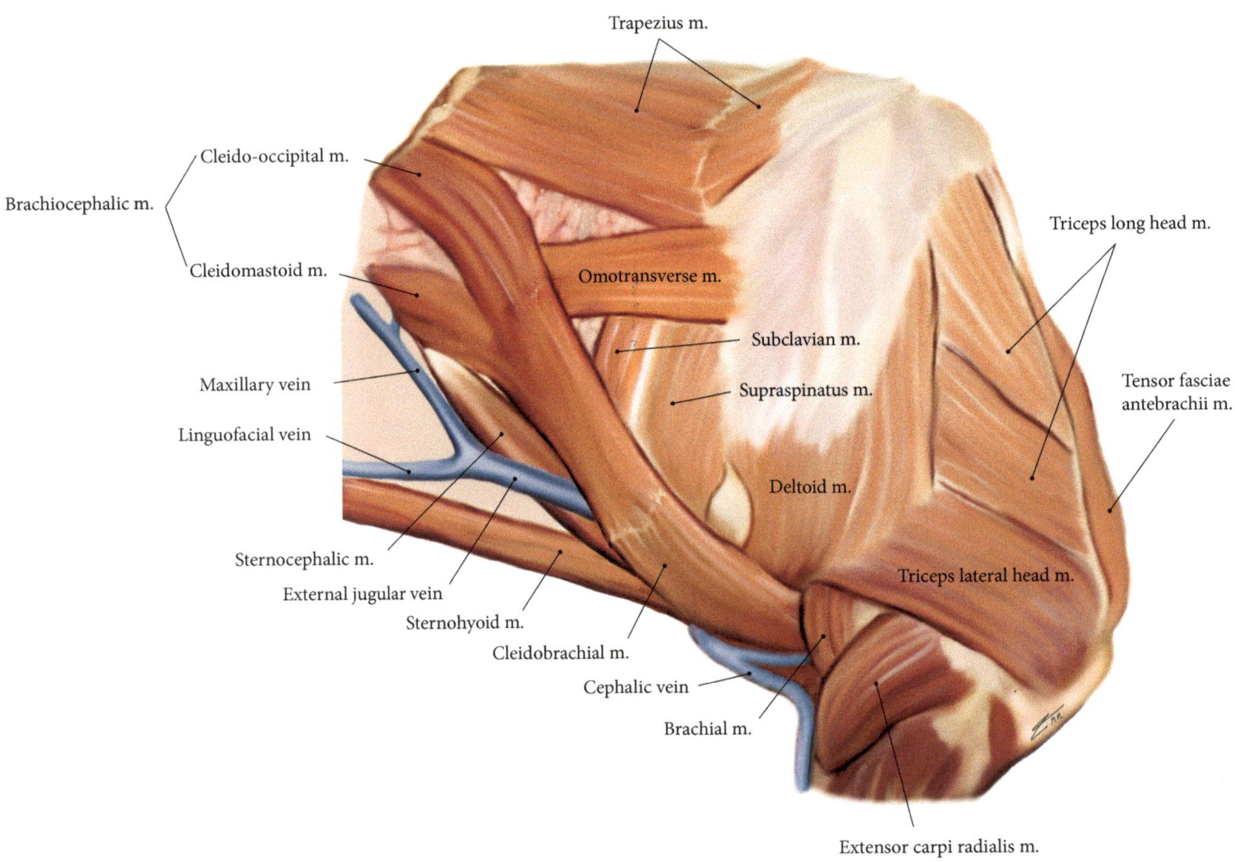

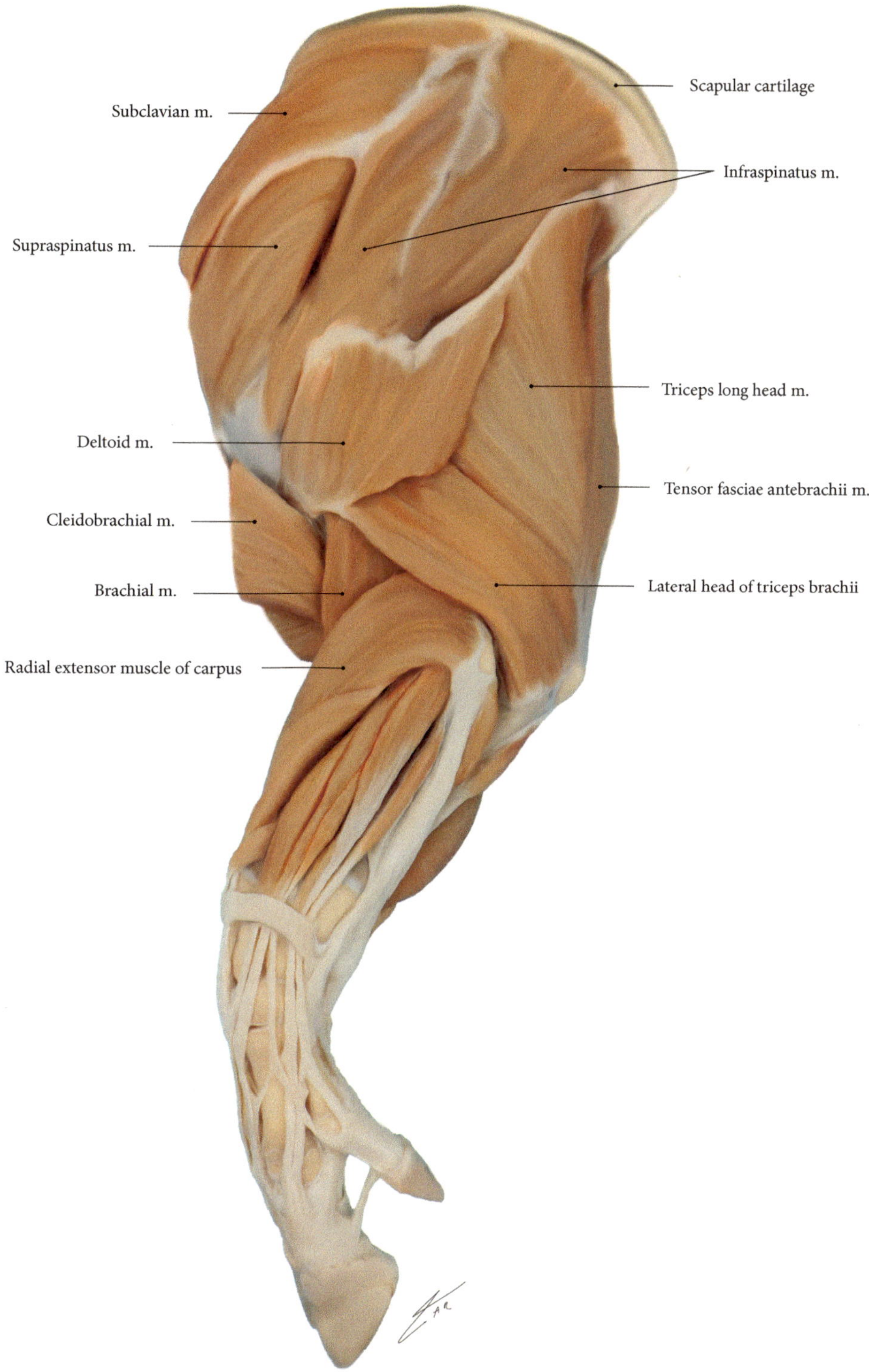
Scapular cartilage
Subclavian m.
Infraspinatus m.
Supraspinatus m.
Triceps long head m.
Deltoid m.
Tensor fasciae antebrachii m.
Cleidobrachial m.
Brachial m.
Lateral head of triceps brachii
Radial extensor muscle of carpus

Muscles of the thoracic limb – medial view

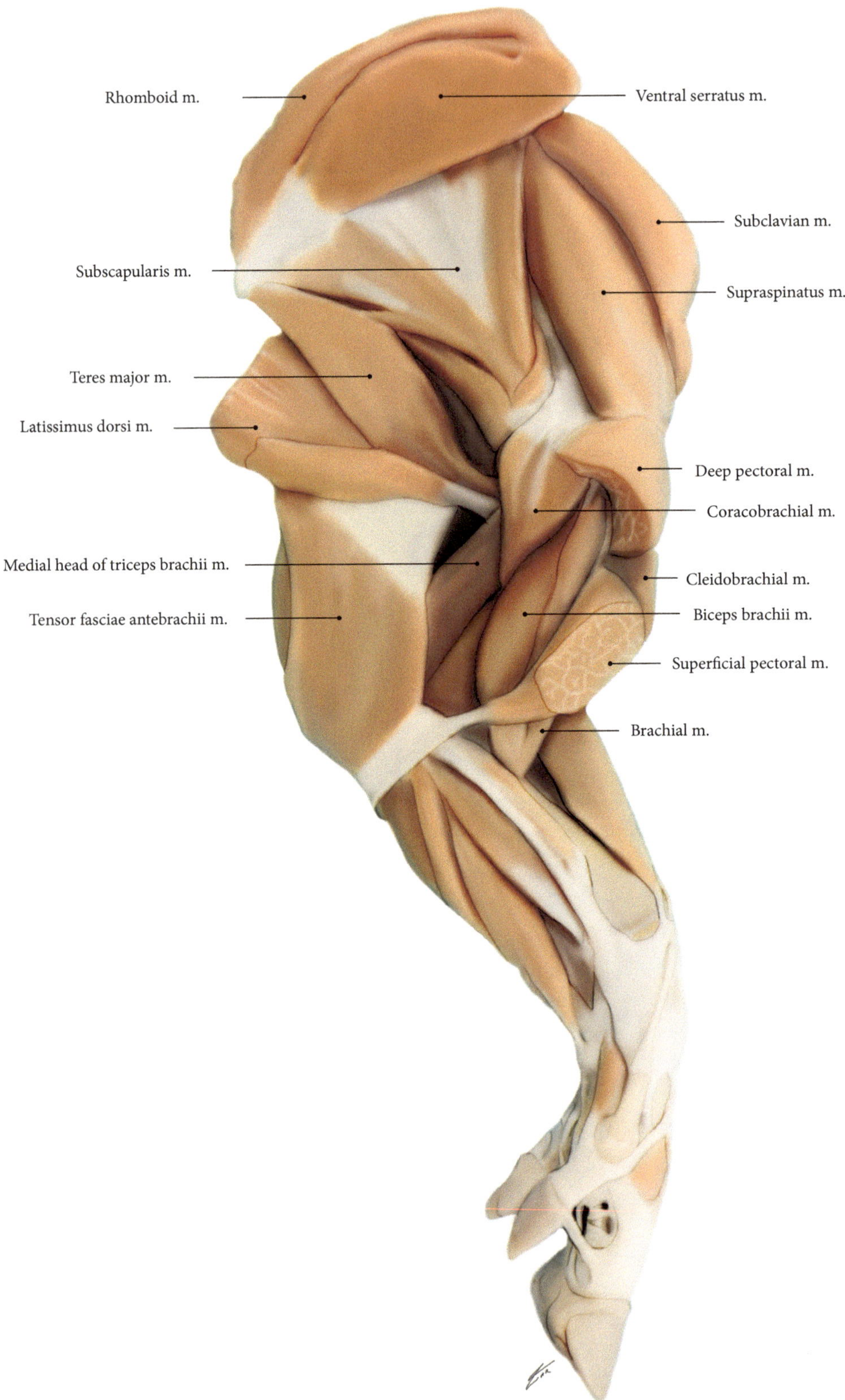

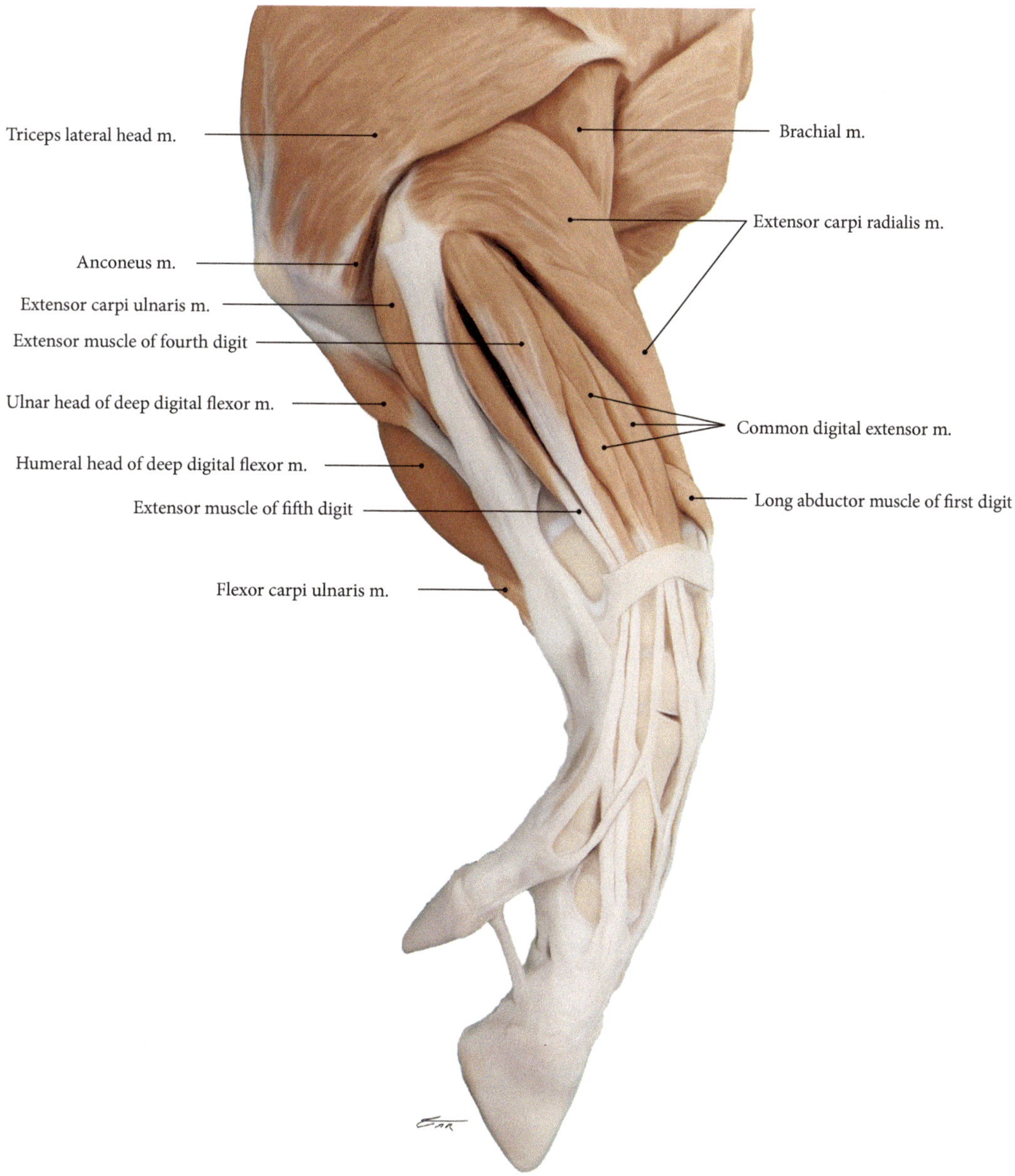
Triceps lateral head m.
Brachial m.
Extensor carpi radialis m.
Anconeus m.
Extensor carpi ulnaris m.
Extensor muscle of fourth digit
Ulnar head of deep digital flexor m.
Common digital extensor m.
Humeral head of deep digital flexor m.
Long abductor muscle of first digit
Extensor muscle of fifth digit
Flexor carpi ulnaris m.

Muscles of the thoracic limb – medial view

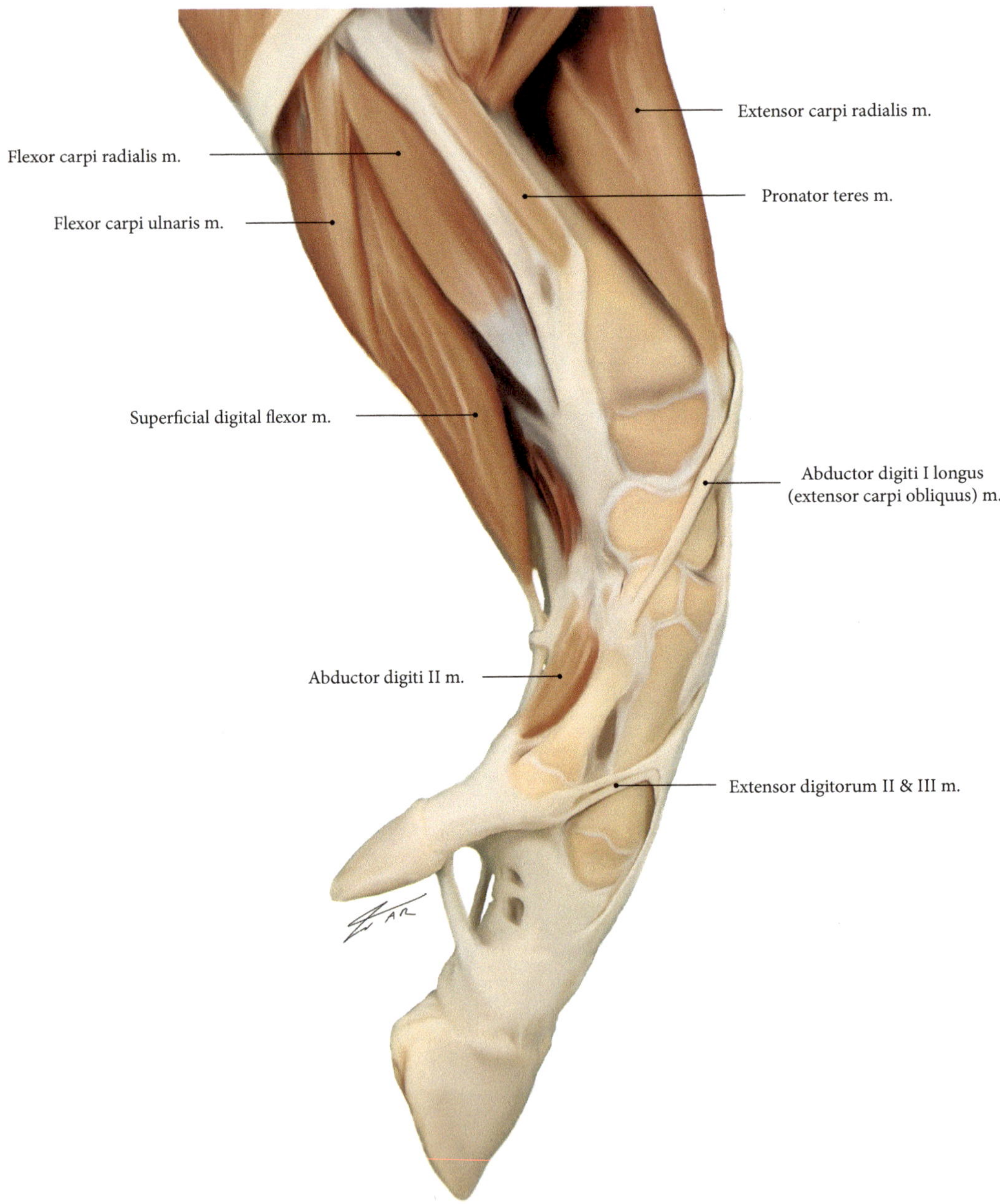

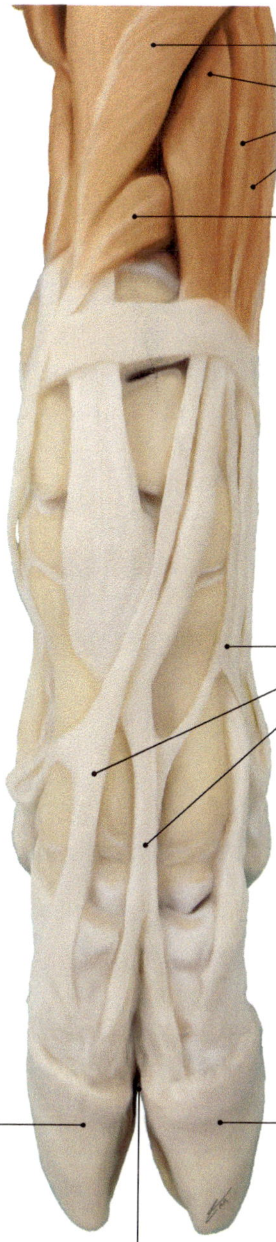
Extensor carpi radialis m.
Common digital extensor m.
Long abductor muscle of first digit
Medial
Lateral
Common digital extensor m.
Digit IV
Digit III
Interdigital cleft

Muscles of the thoracic limb – palmar view

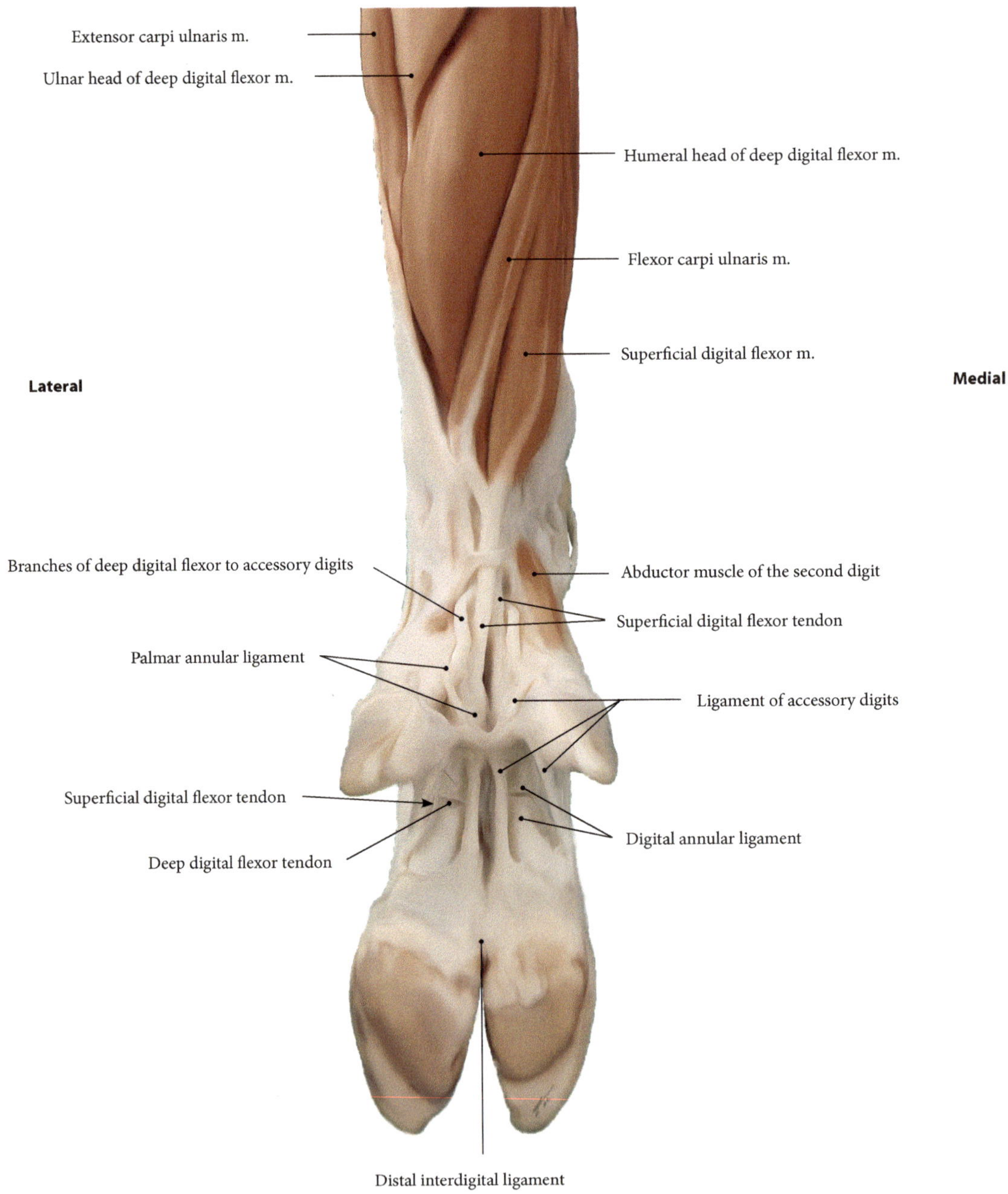

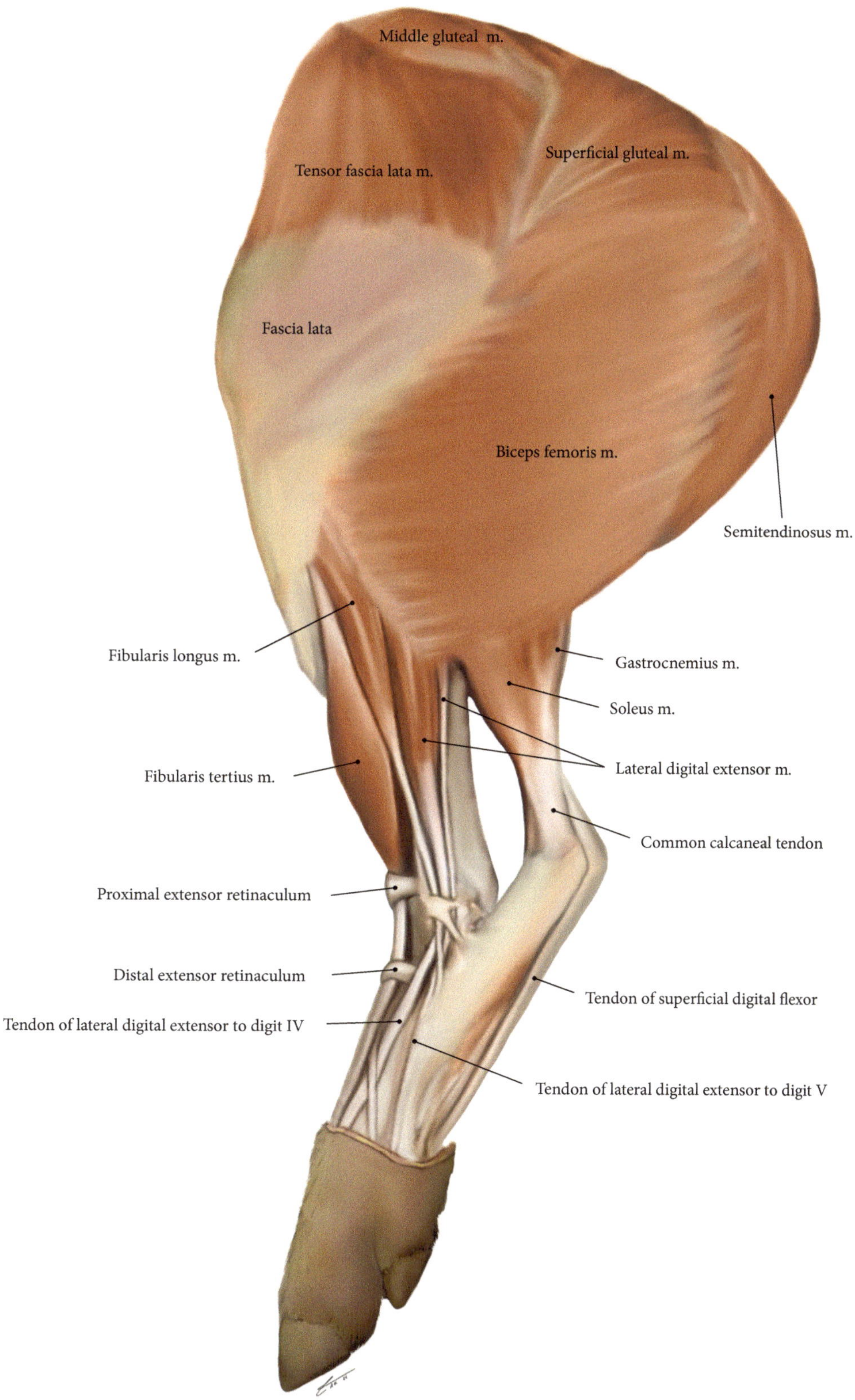
Middle gluteal m.
Tensor fascia lata m.
Superficial gluteal m.
Fascia lata
Biceps femoris m.
Semitendinosus m.
Fibularis longus m.
Gastrocnemius m.
Soleus m.
Fibularis tertius m.
Lateral digital extensor m.
Common calcaneal tendon
Proximal extensor retinaculum
Distal extensor retinaculum
Tendon of superficial digital flexor
Tendon of lateral digital extensor to digit IV
Tendon of lateral digital extensor to digit V

Muscles of the pelvic limb – medial view

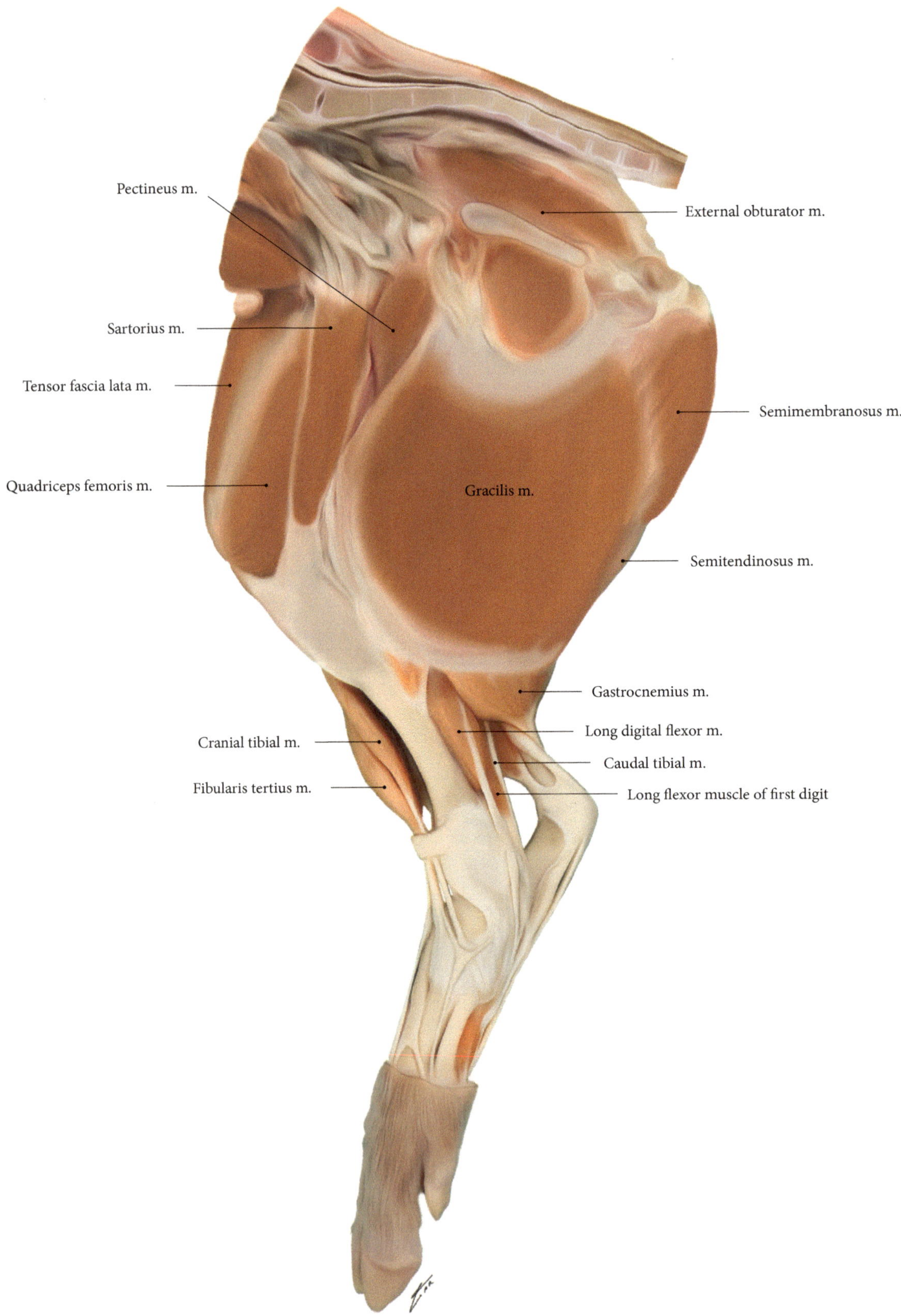

Muscles of the distal pelvic limb – lateral view

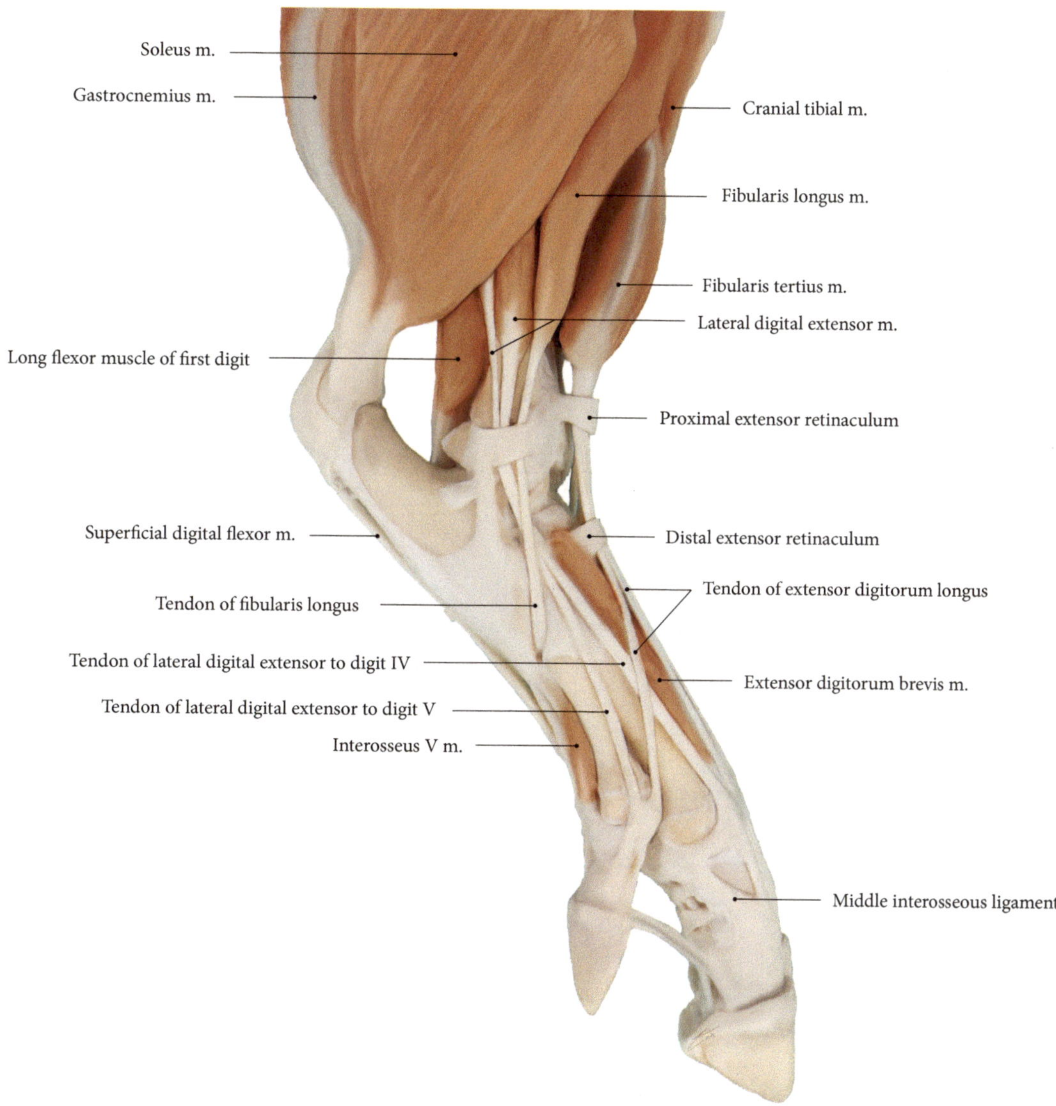

Muscles of the distal pelvic limb – medial view

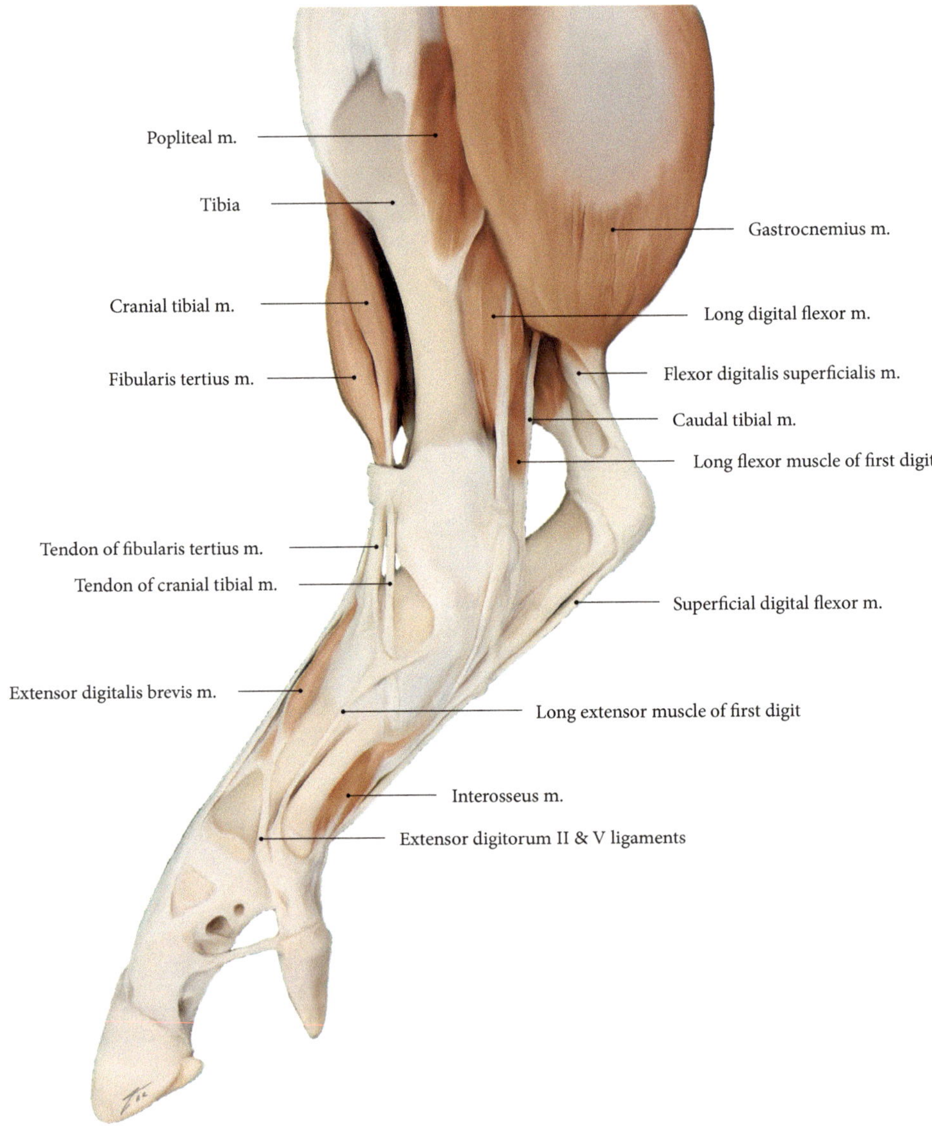

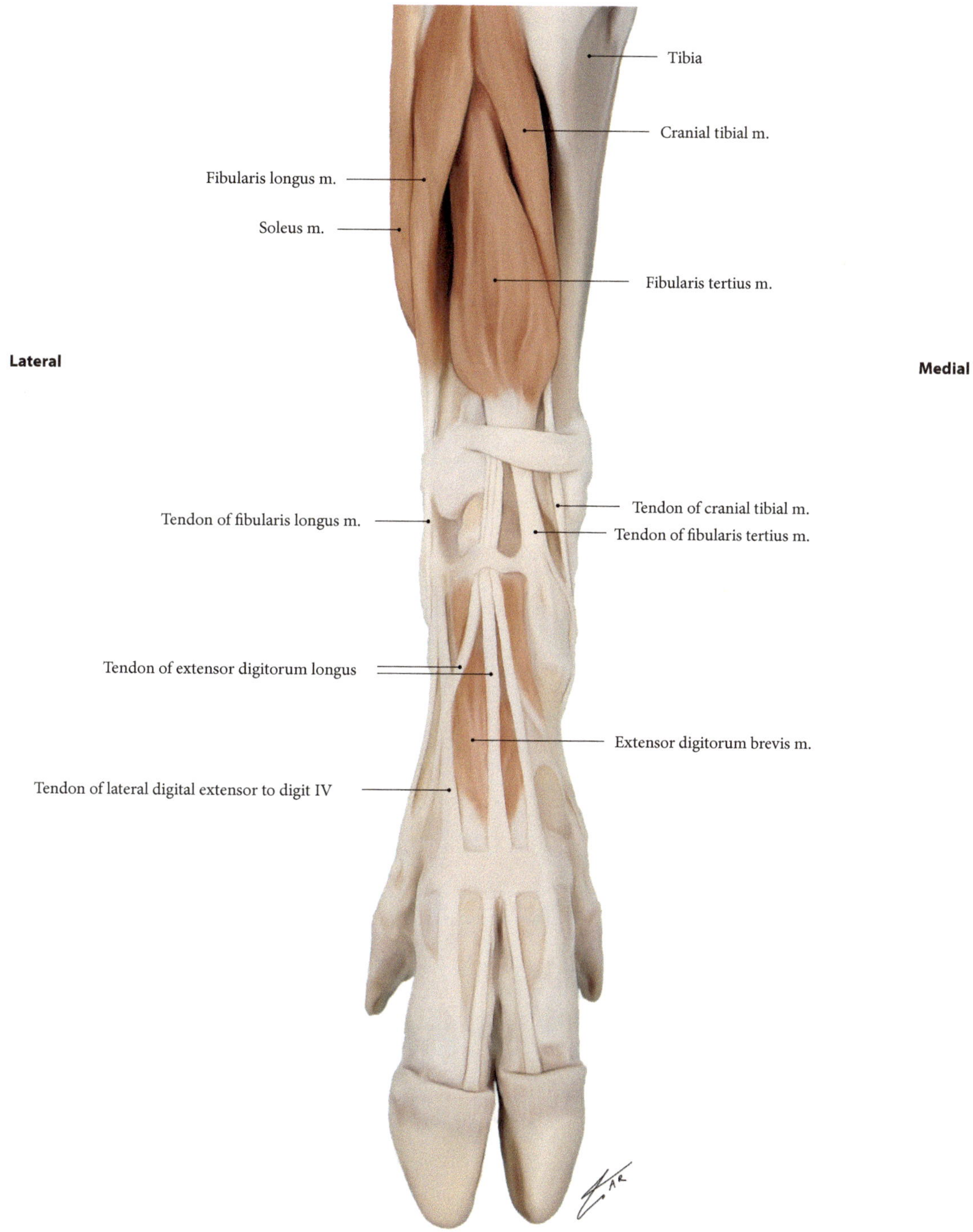
Tibia
Cranial tibial m.
Fibularis longus m.
Soleus m.
Fibularis tertius m.
Lateral
Medial
Tendon of cranial tibial m.
Tendon of fibularis longus m.
Tendon of fibularis tertius m.
Tendon of extensor digitorum longus
Extensor digitorum brevis m.
Tendon of lateral digital extensor to digit IV

Muscles of the distal pelvic limb – plantar view

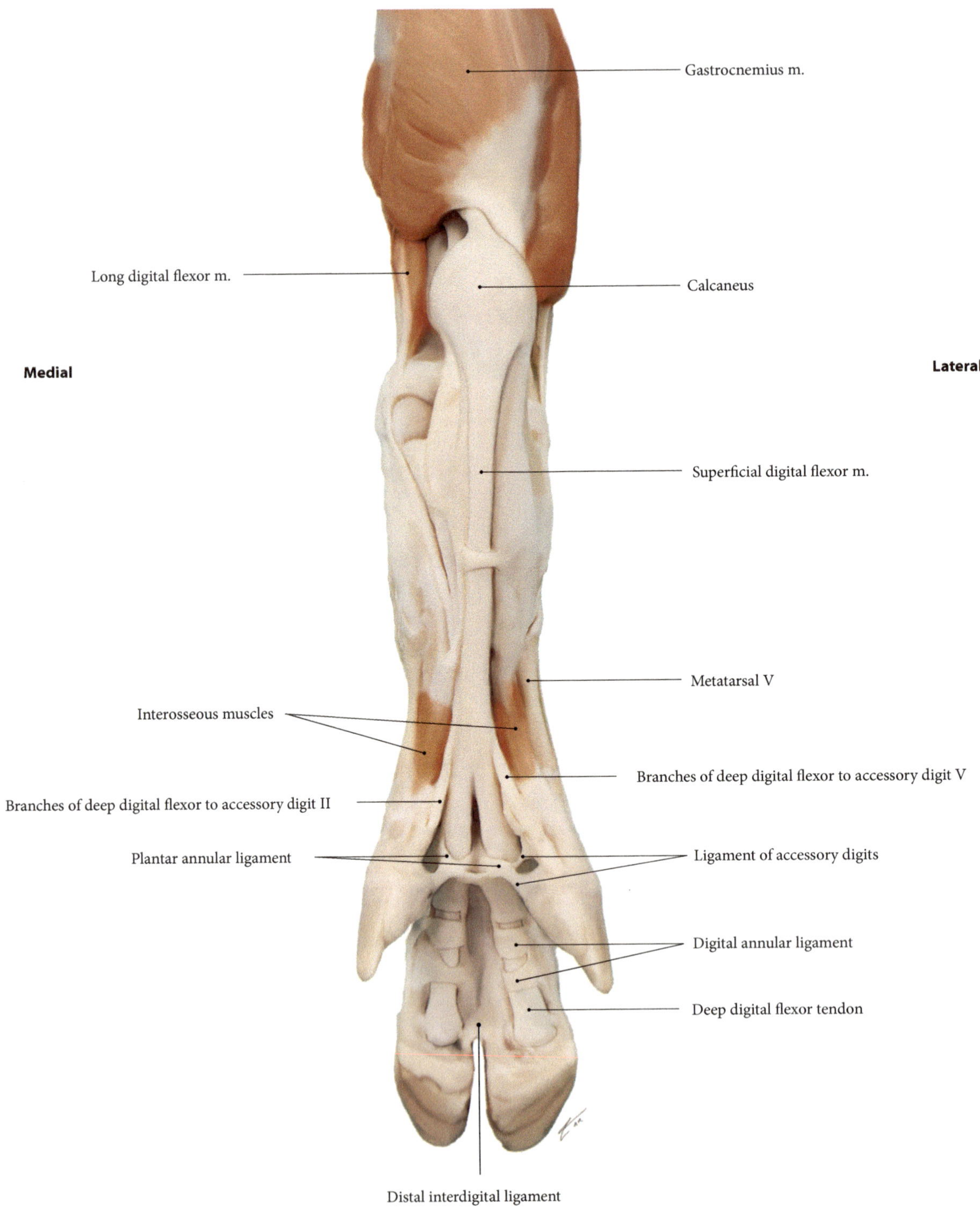

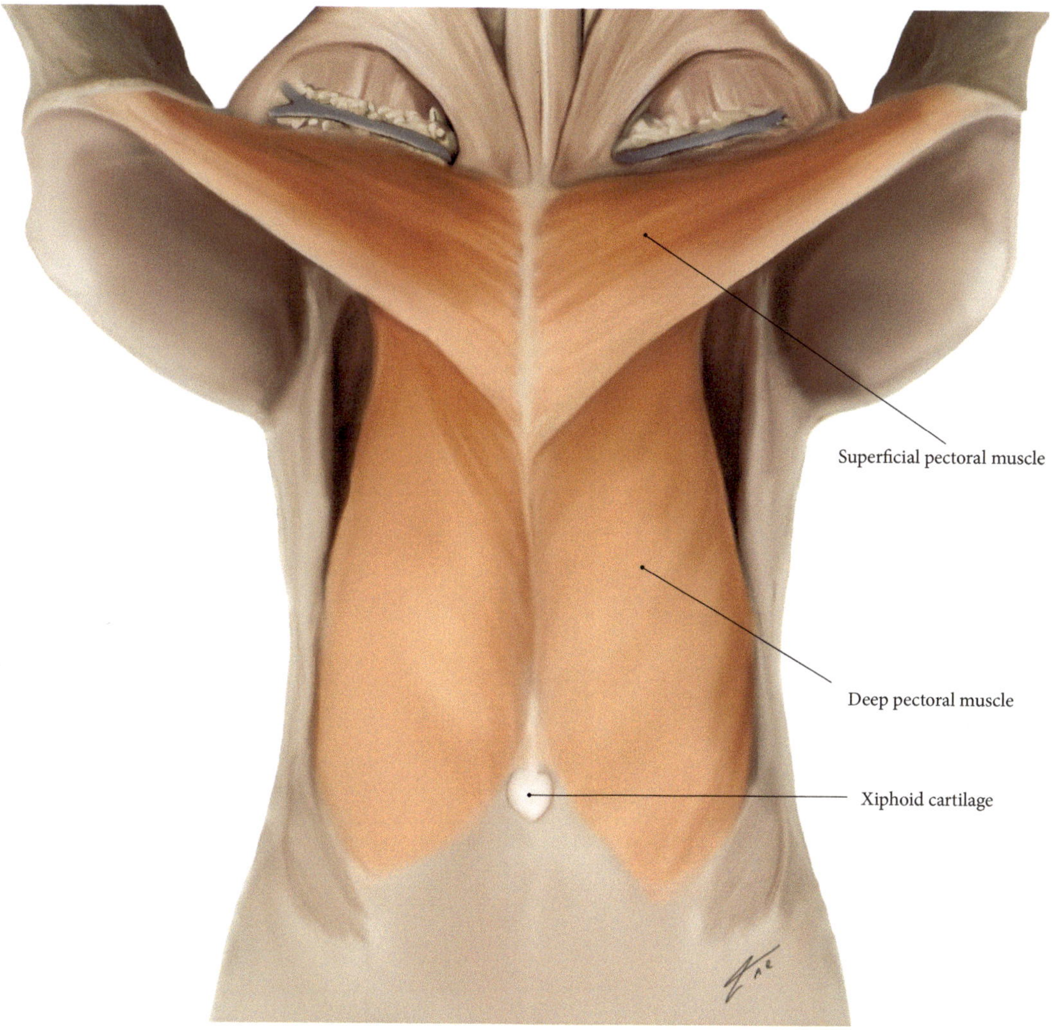
Superficial pectoral muscle
Deep pectoral muscle
Xiphoid cartilage

Deep pectoral muscles – ventral view

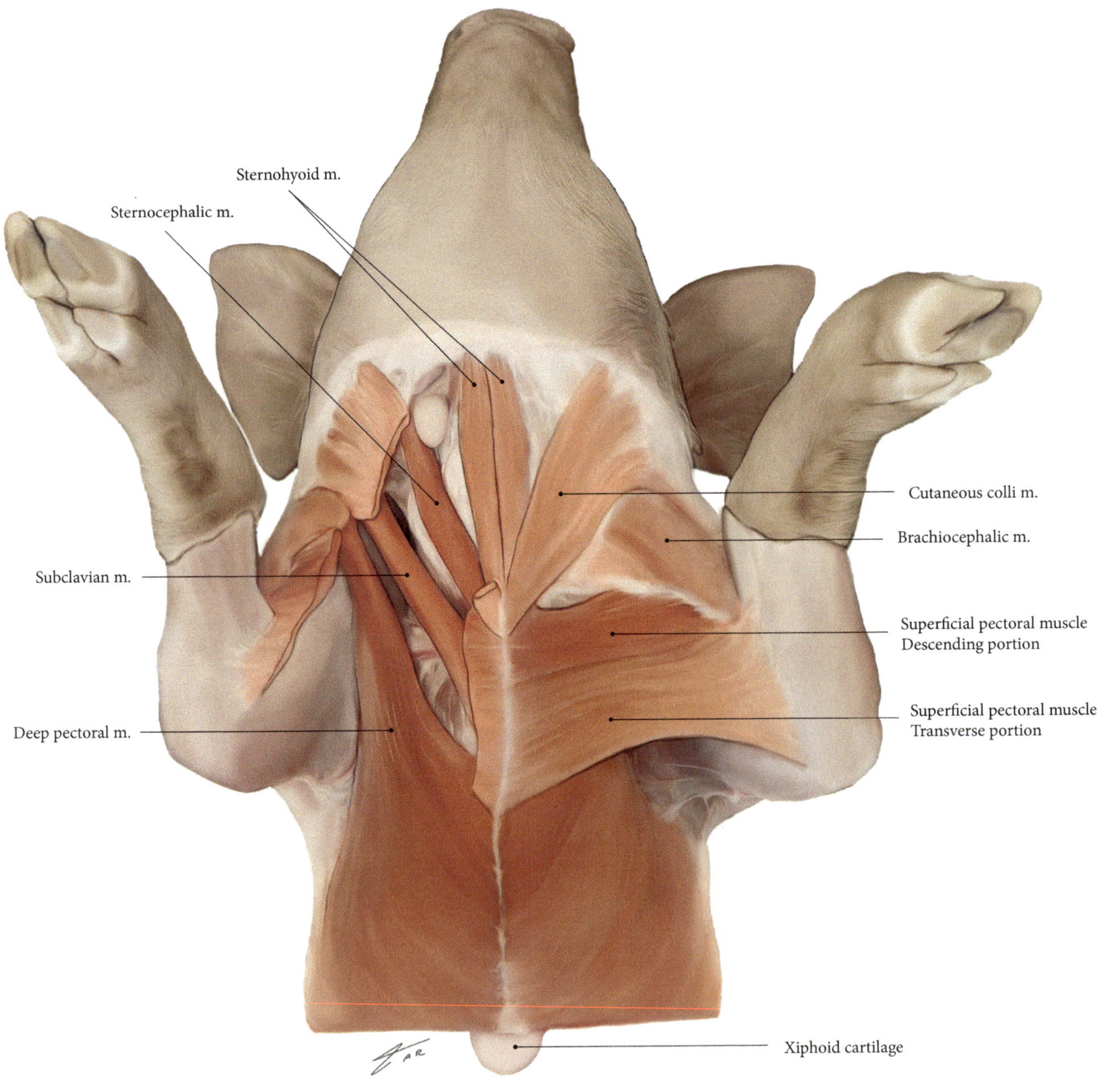

Abdominal muscles – lateral view

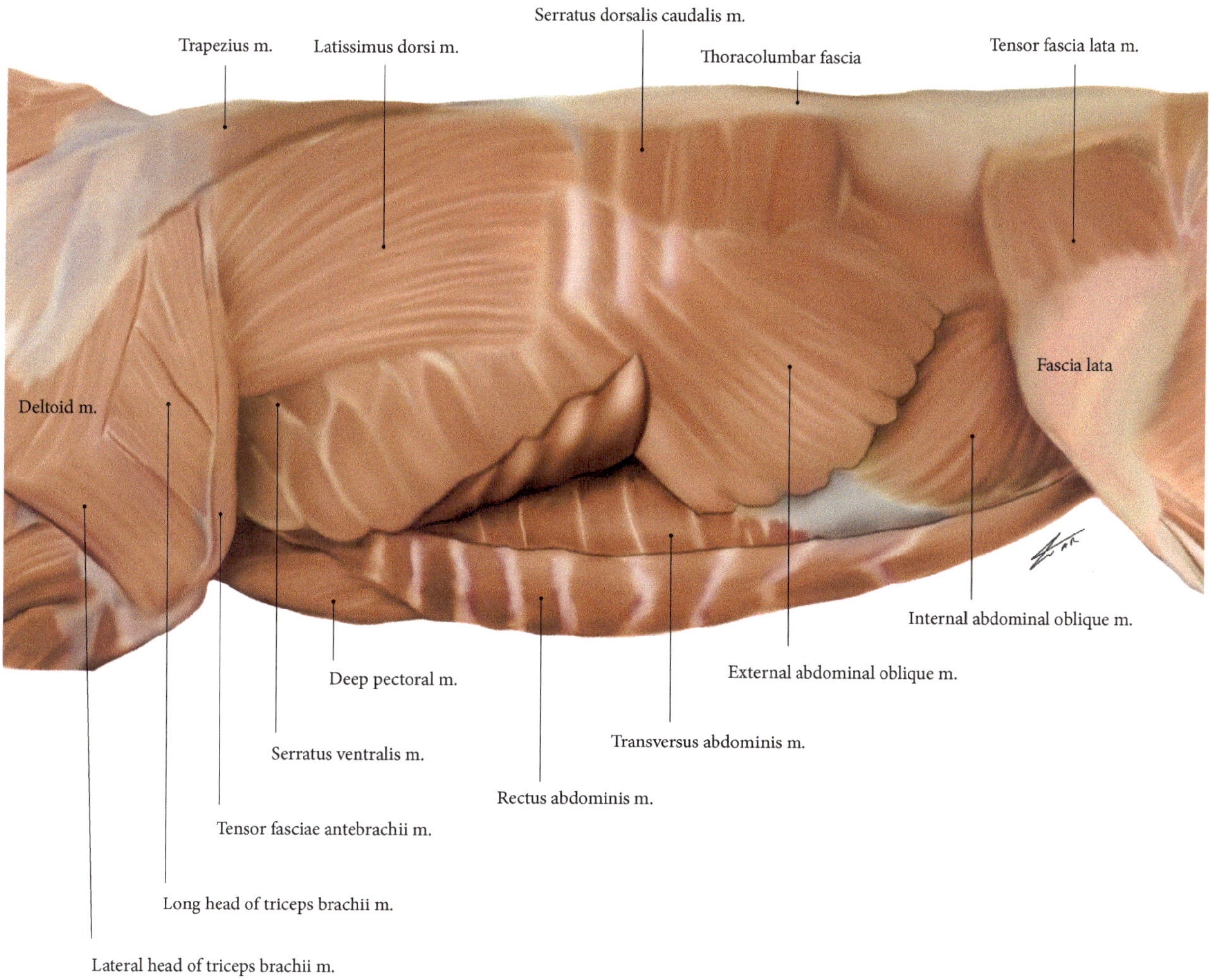

Teeth – lateral aspect within the skull

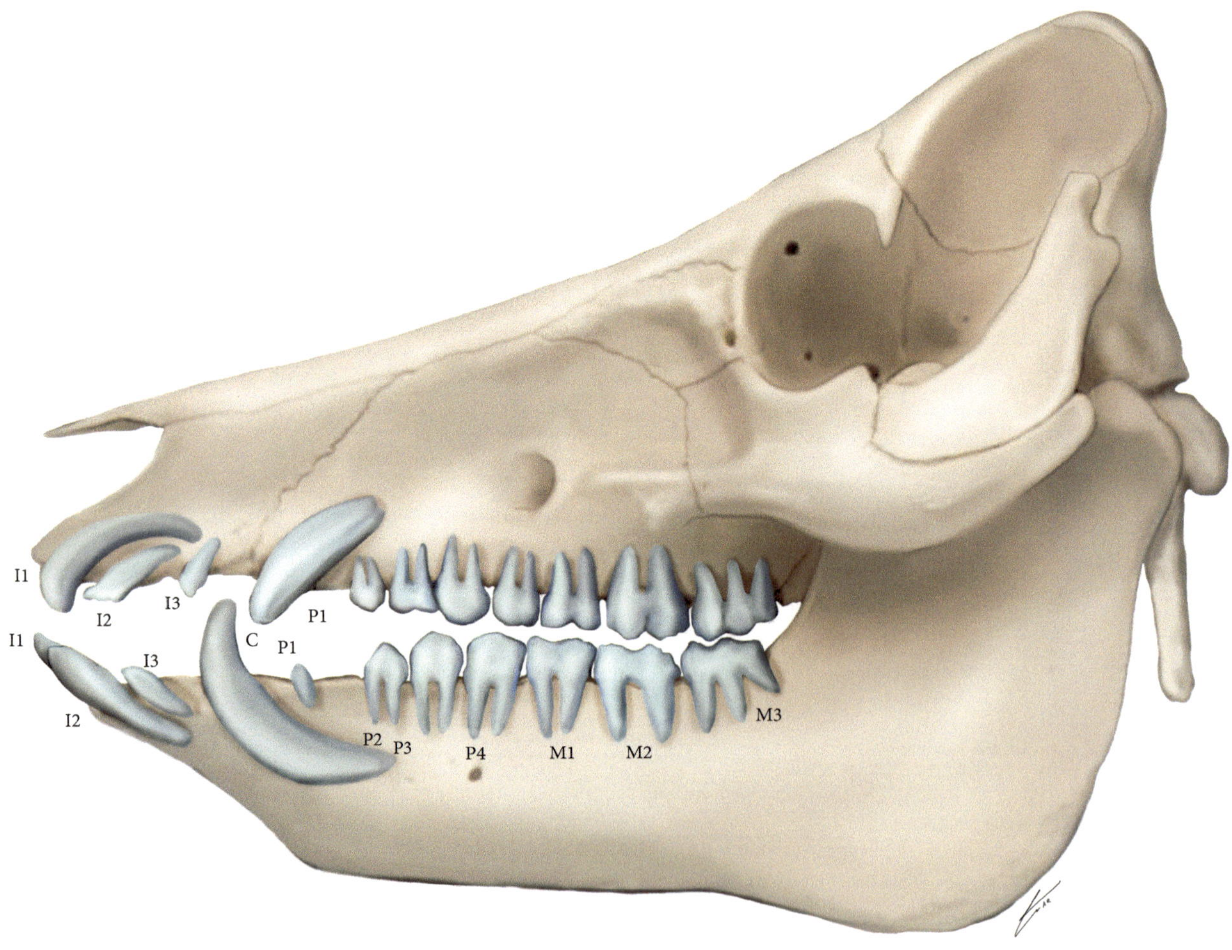

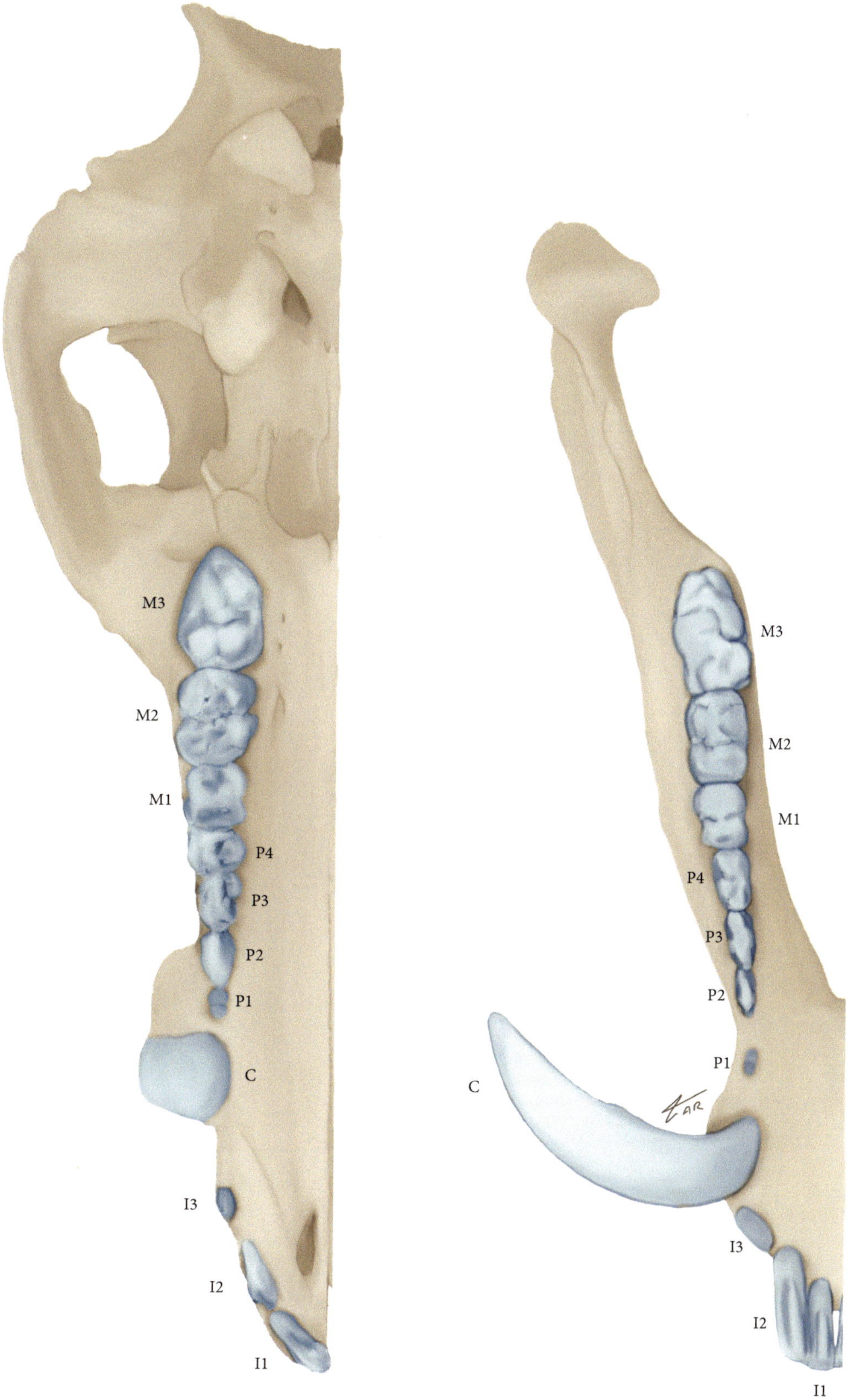
M3
M2
M1
P4
P3
P2
P1
C
I3
I2
I1
M3
M2
M1
P4
P3
P2
P1
C
I3
I2
I1

Salivary glands

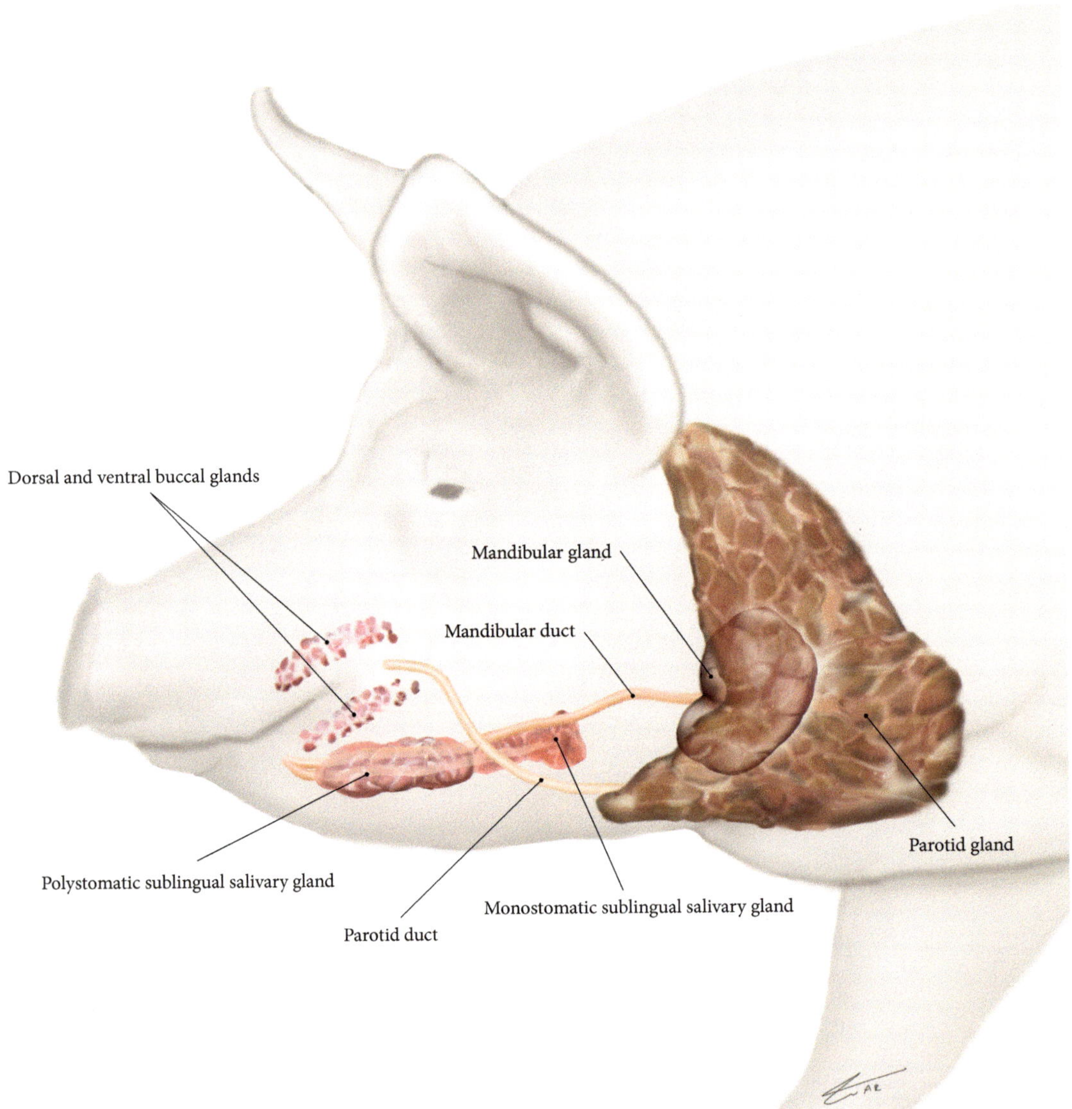

Median section of the porcine head

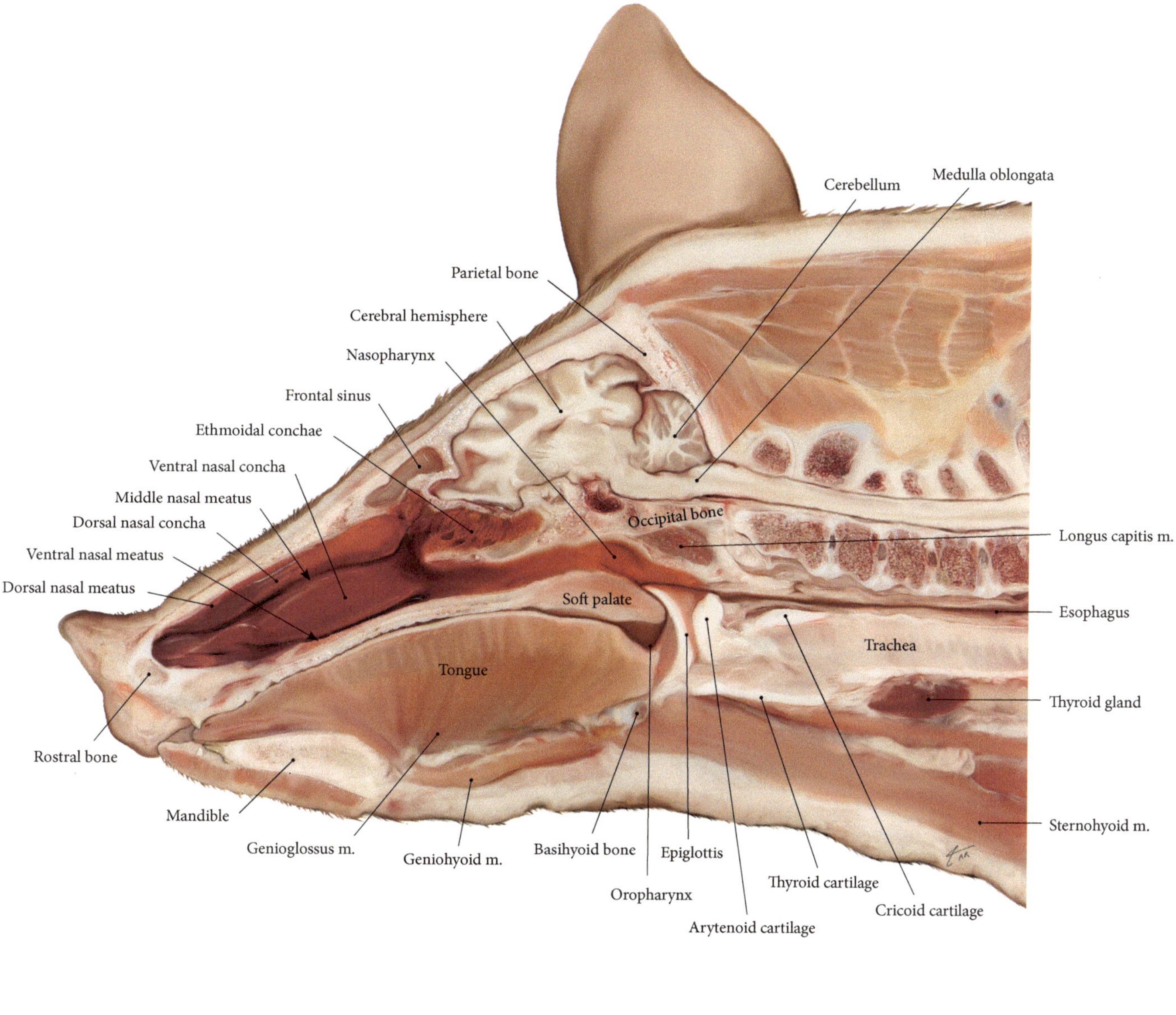

Hard palate – young pig

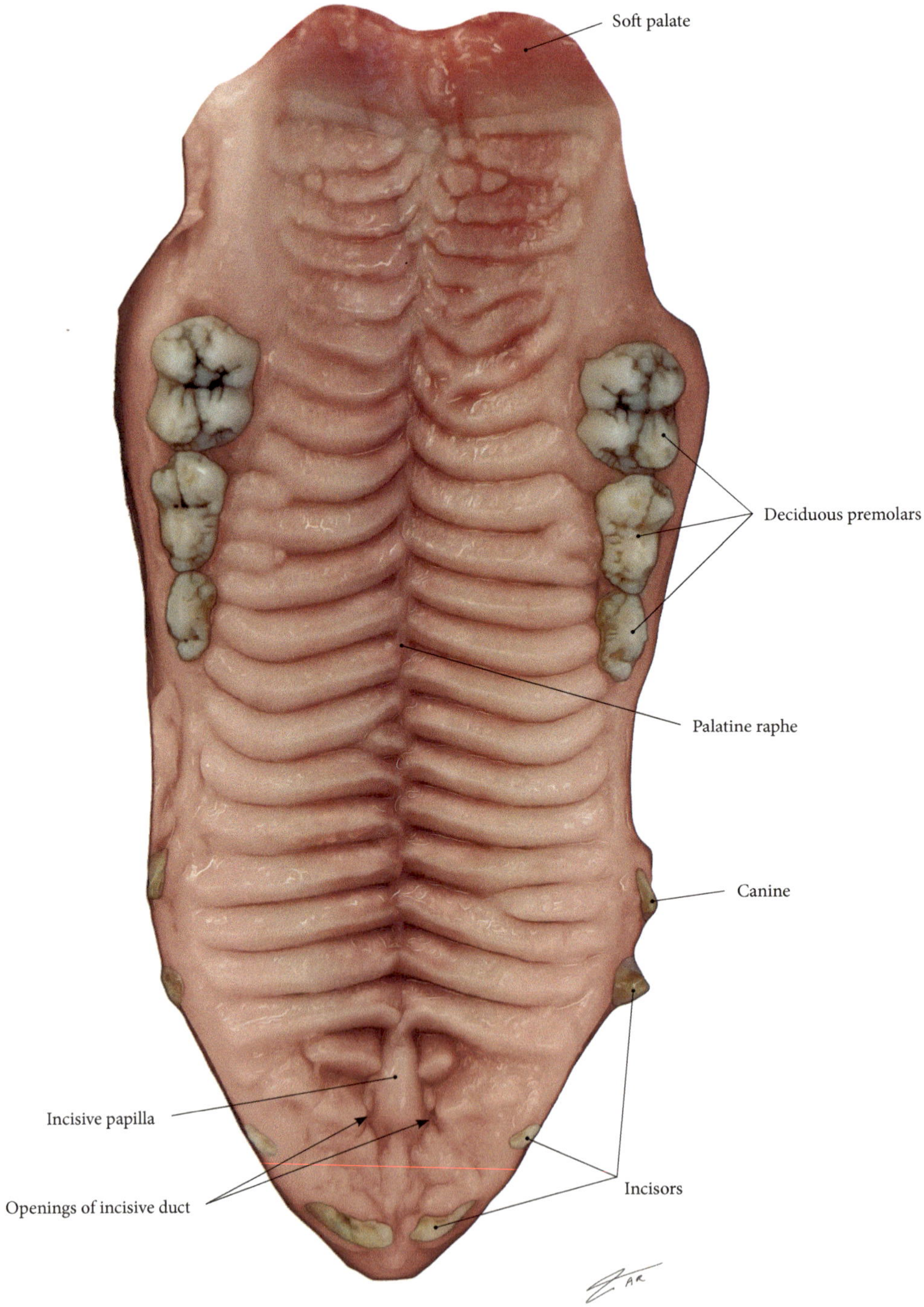

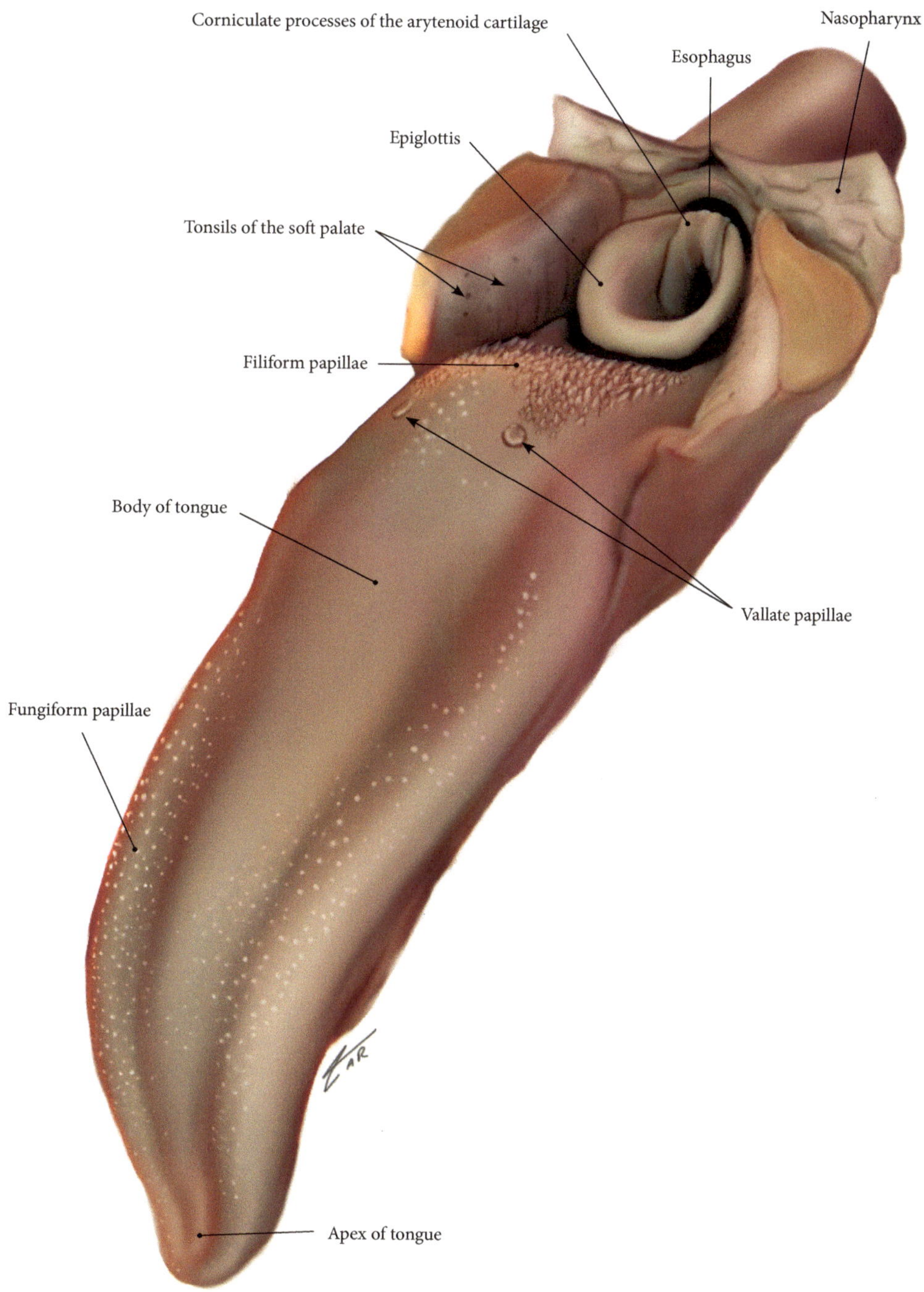
Corniculate processes of the arytenoid cartilage
Nasopharynx
Esophagus
Epiglottis
Tonsils of the soft palate
Filiform papillae
Body of tongue
Vallate papillae
Fungiform papillae
Apex of tongue

Esophagus – layers

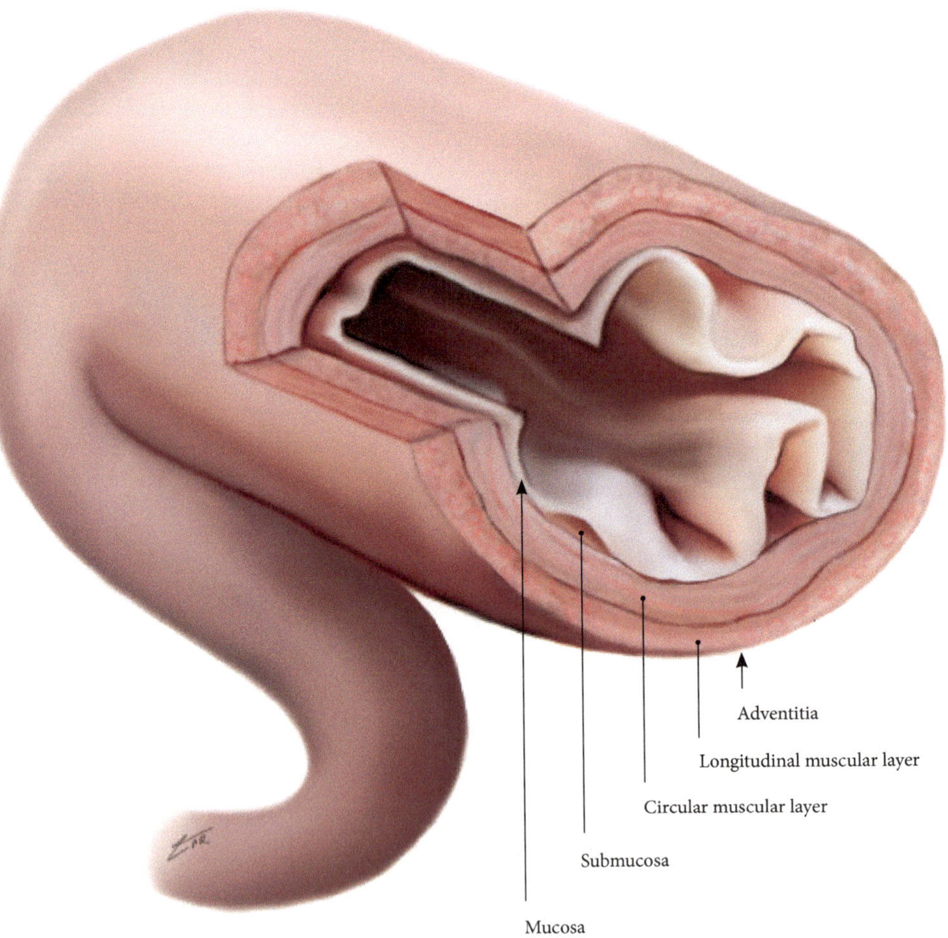

Viscera of the male pig – left side

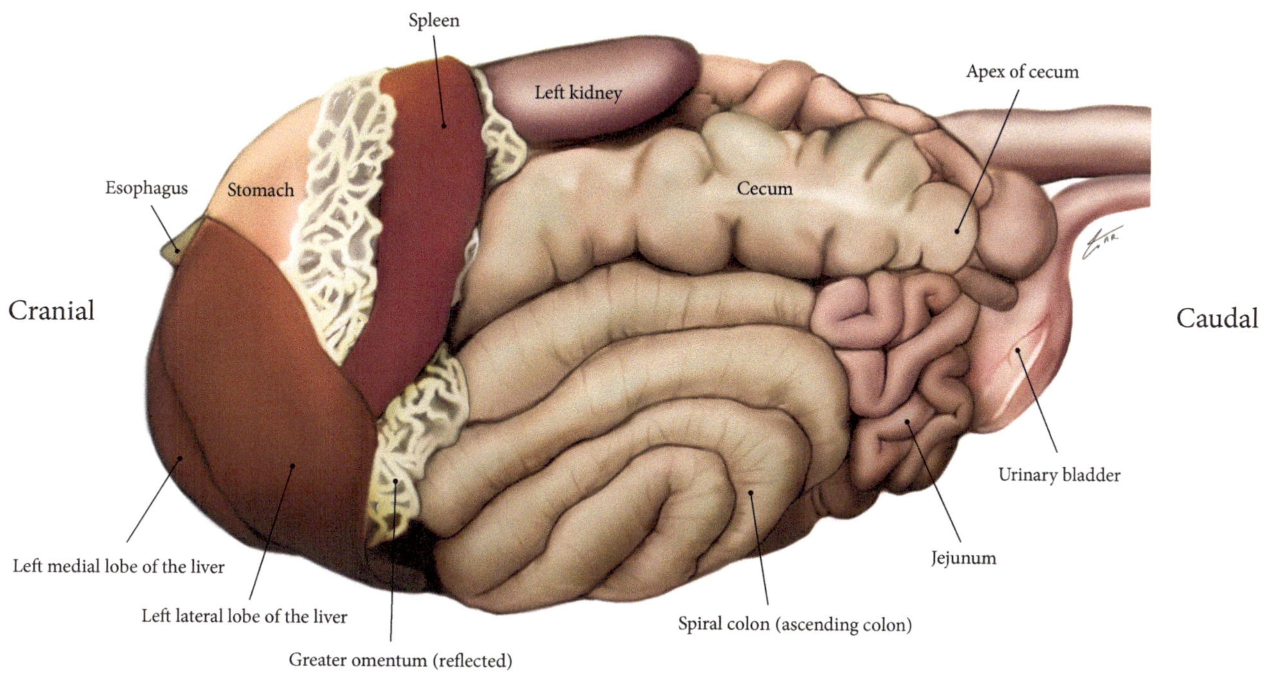

Viscera of the male pig – right side

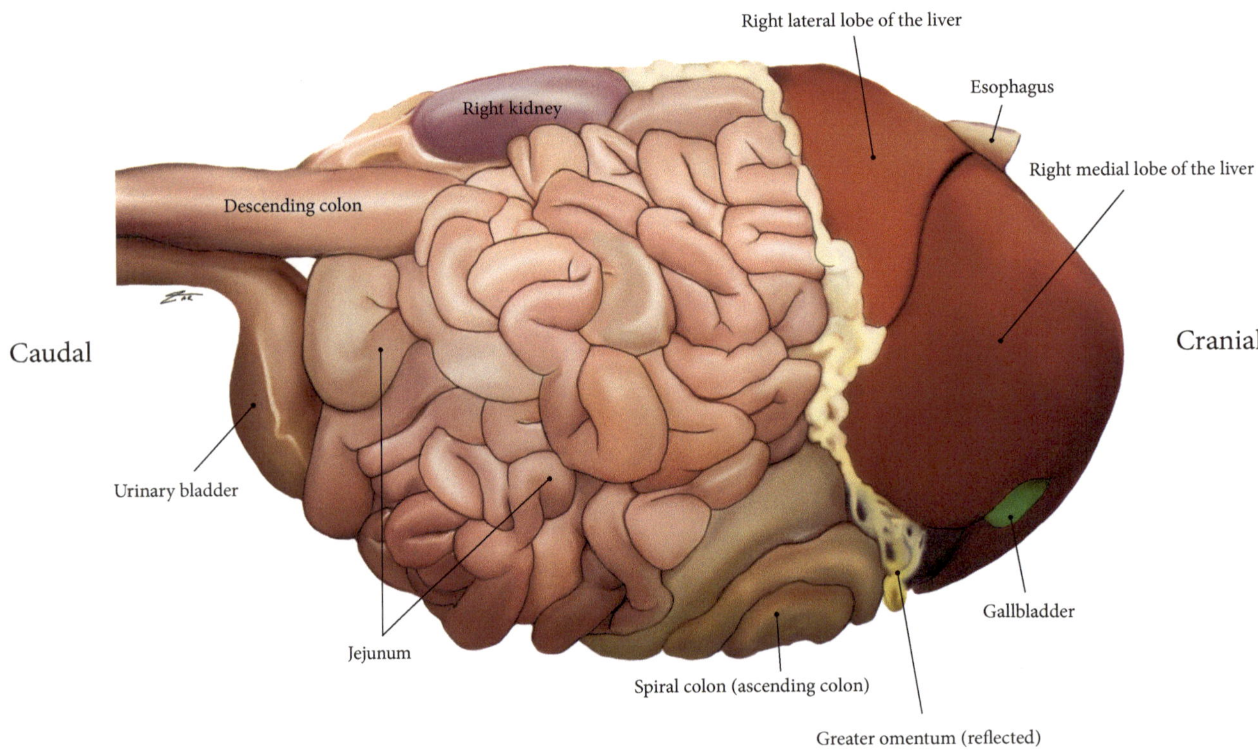

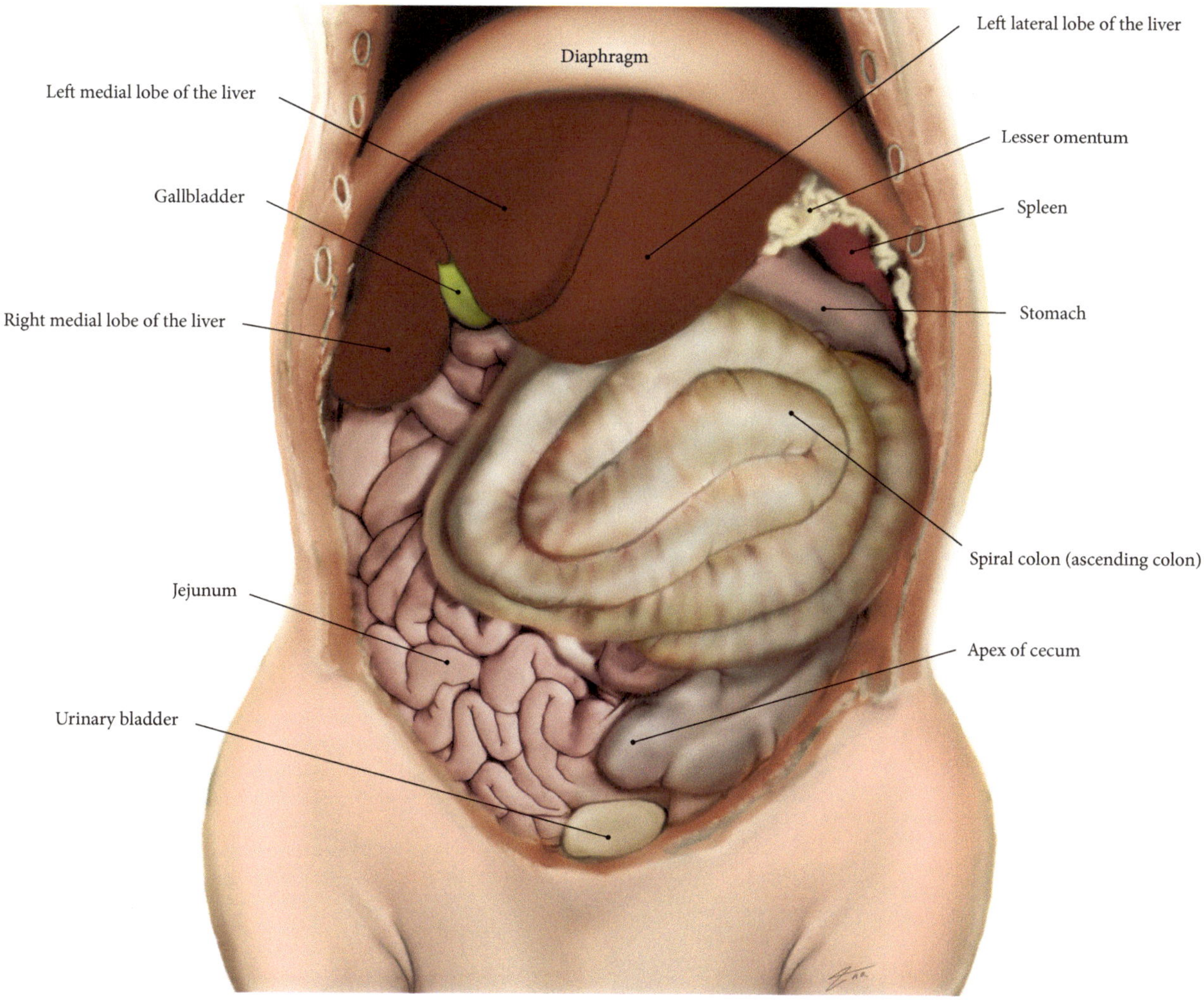
Left lateral lobe of the liver
Diaphragm
Left medial lobe of the liver
Lesser omentum
Gallbladder
Spleen
Right medial lobe of the liver
Stomach
Spiral colon (ascending colon)
Jejunum
Apex of cecum
Urinary bladder

Stomach and duodenum

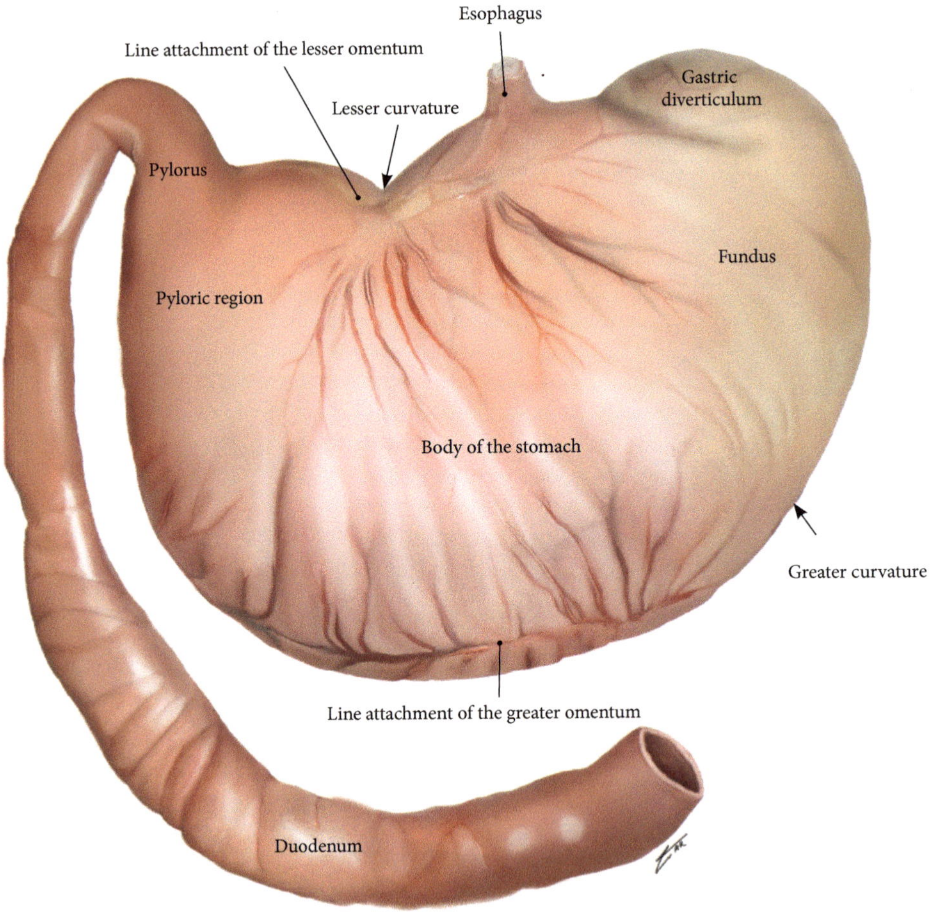

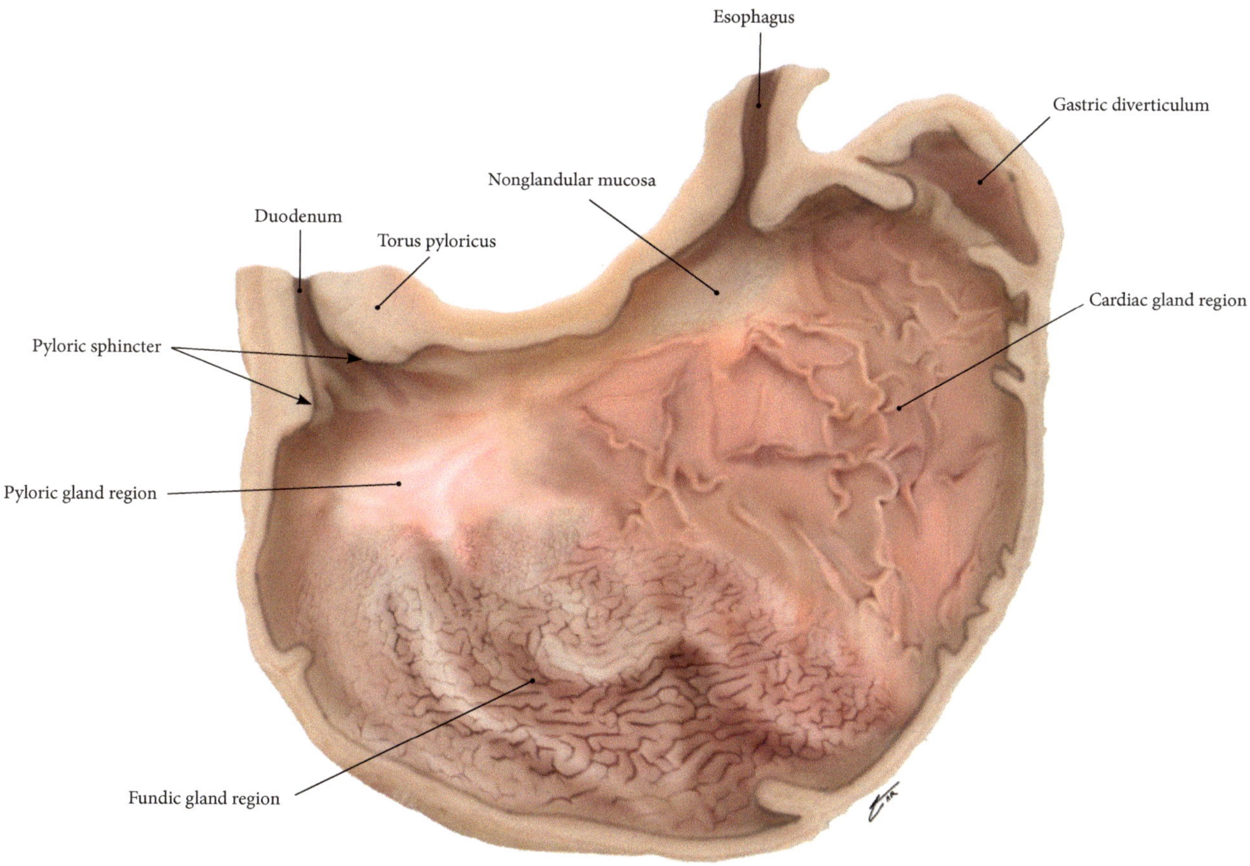
Esophagus
Gastric diverticulum
Nonglandular mucosa
Duodenum
Torus pyloricus
Cardiac gland region
Pyloric sphincter
Pyloric gland region
Fundic gland region

Colon

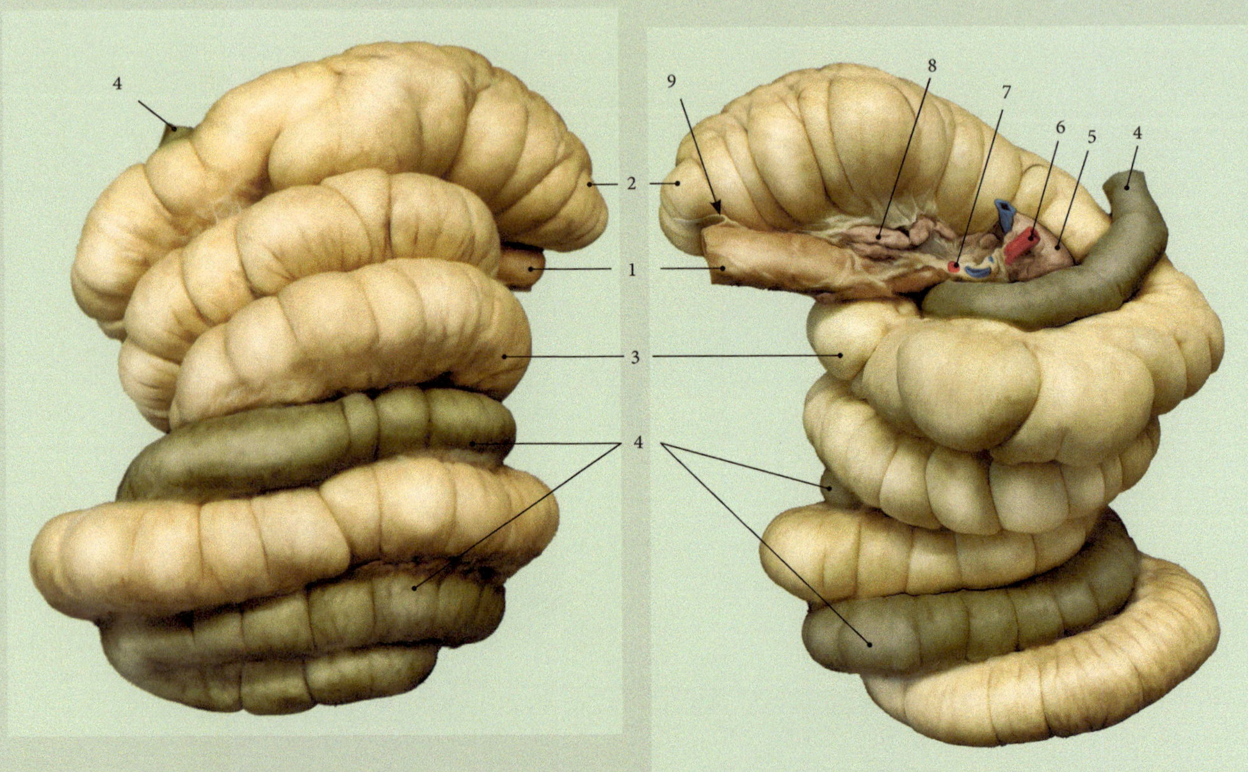

1. Ileum
2. Cecum
3. Centripetal coils of spiral loop of ascending colon
4. Centrifugal coils of spiral loop of ascending colon
5. Colic lymph nodes
6. Right colic artery
7. Ileocecocolic a.
8. Ileocolic lymph nodes
9. Ileocecal ligament

Spiral colon

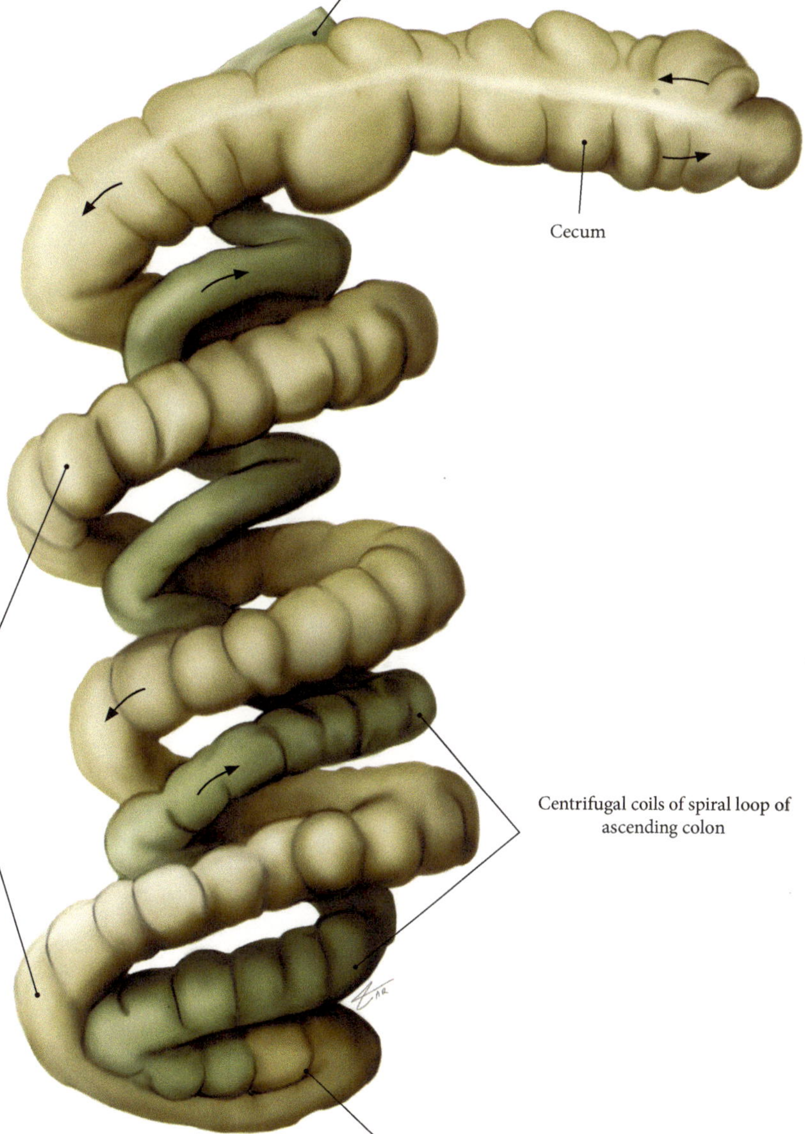

Cecum

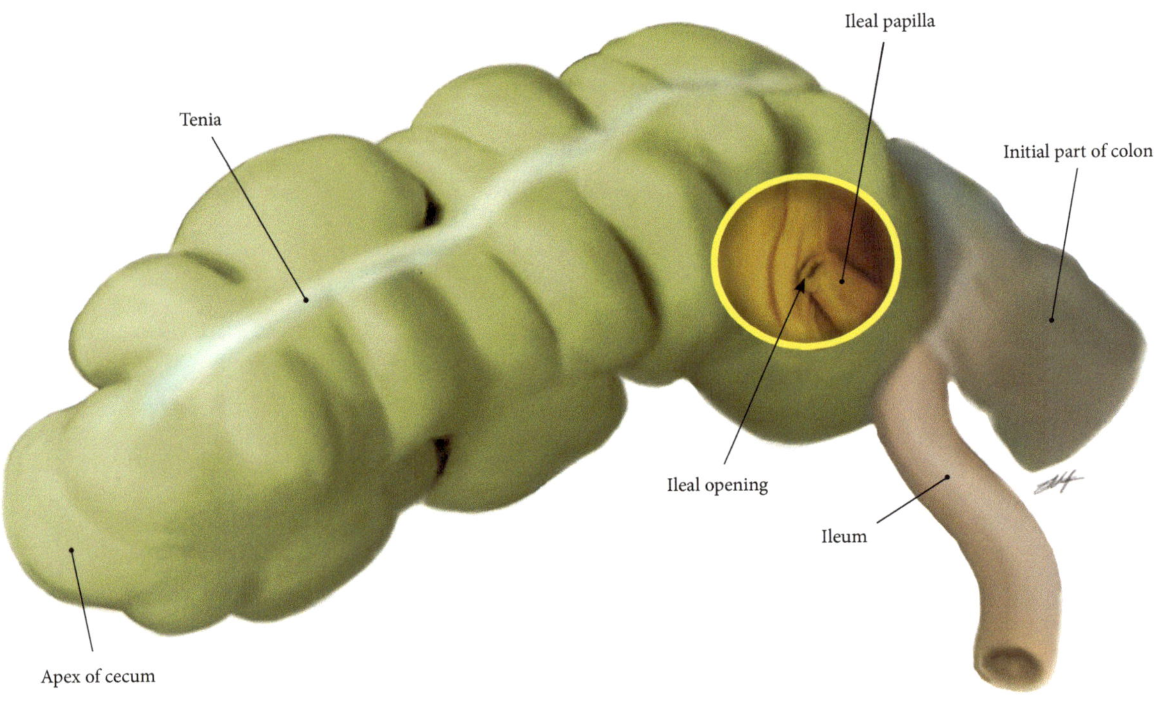
Ileal papilla
Tenia
Initial part of colon
Ileal opening
Ileum
Apex of cecum

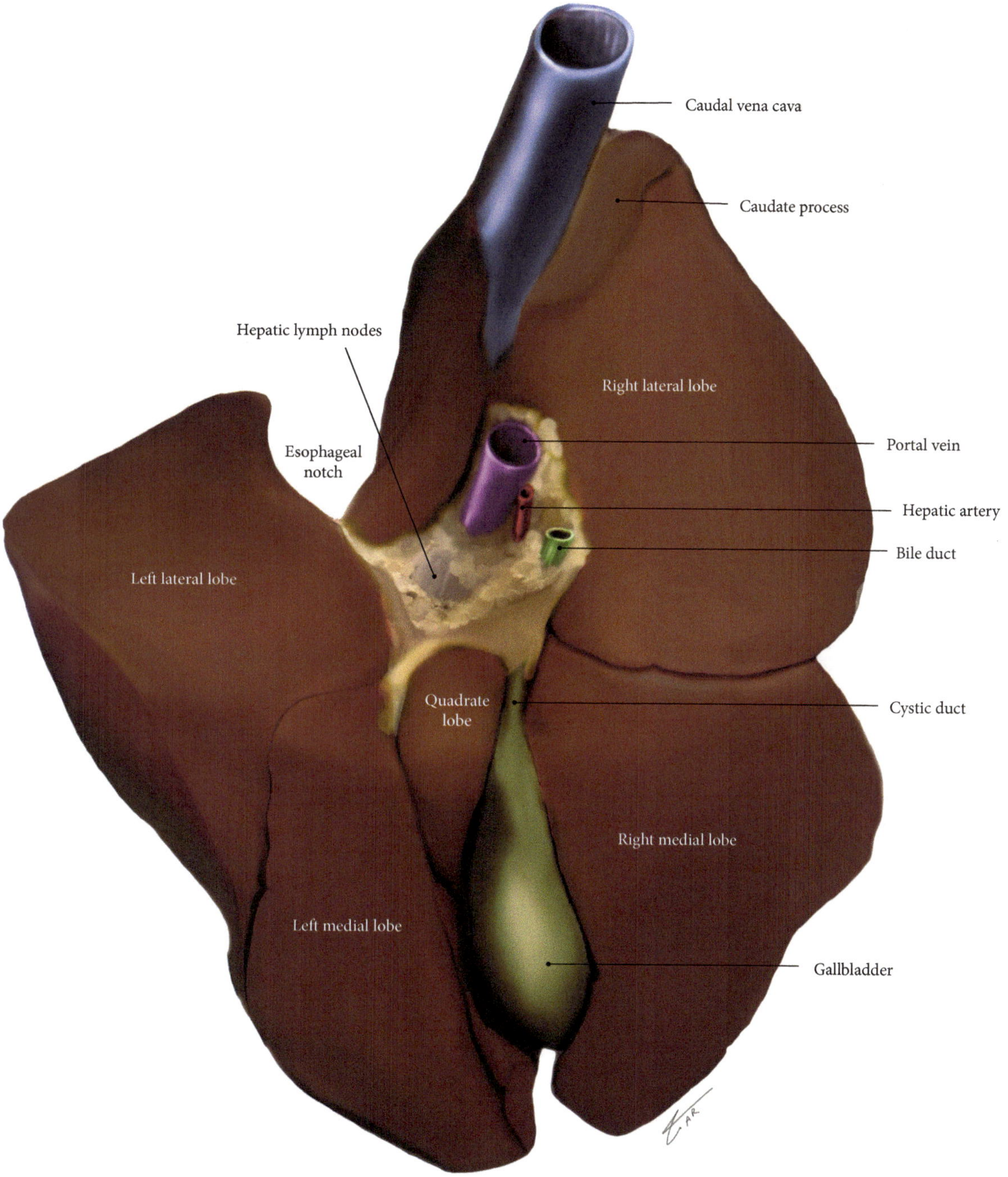
Caudal vena cava
Caudate process
Hepatic lymph nodes
Right lateral lobe
Esophageal
notch
Portal vein
Hepatic artery
Bile duct
Left lateral lobe
Quadrate
lobe
Cystic duct
Right medial lobe
Left medial lobe
Gallbladder

Liver of pig – parietal surface

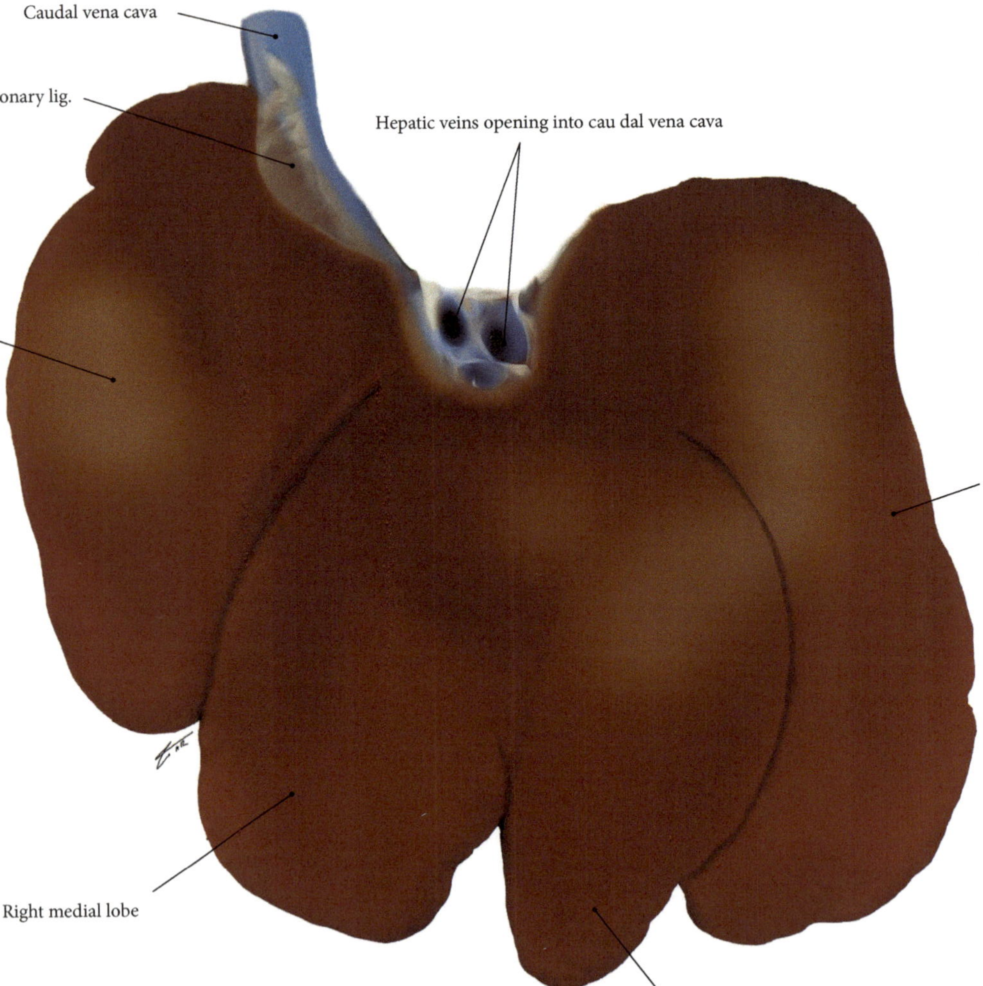

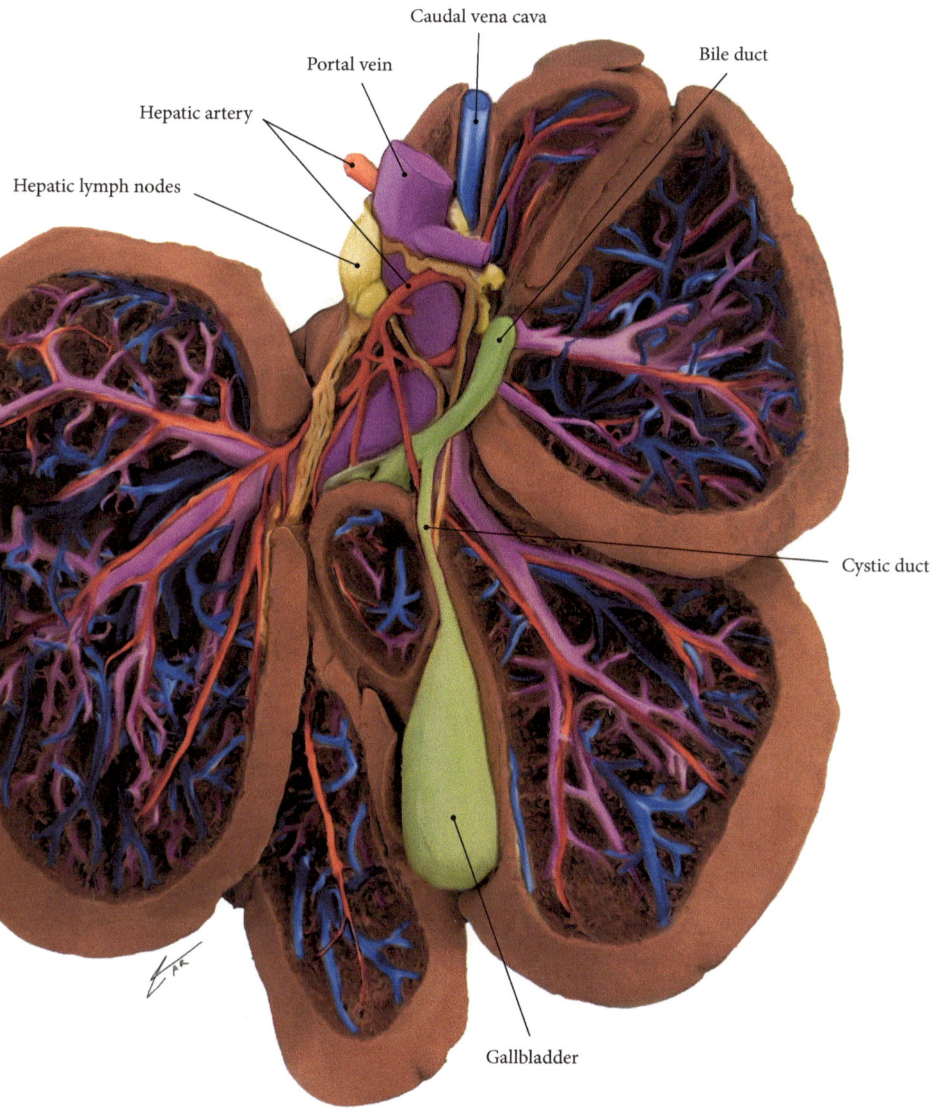

Color coding

- Portal veins
- Hepatic veins
- Hepatic artery
- Biliary system
- Hepatic lymphatic system

Portal vein – visceral view

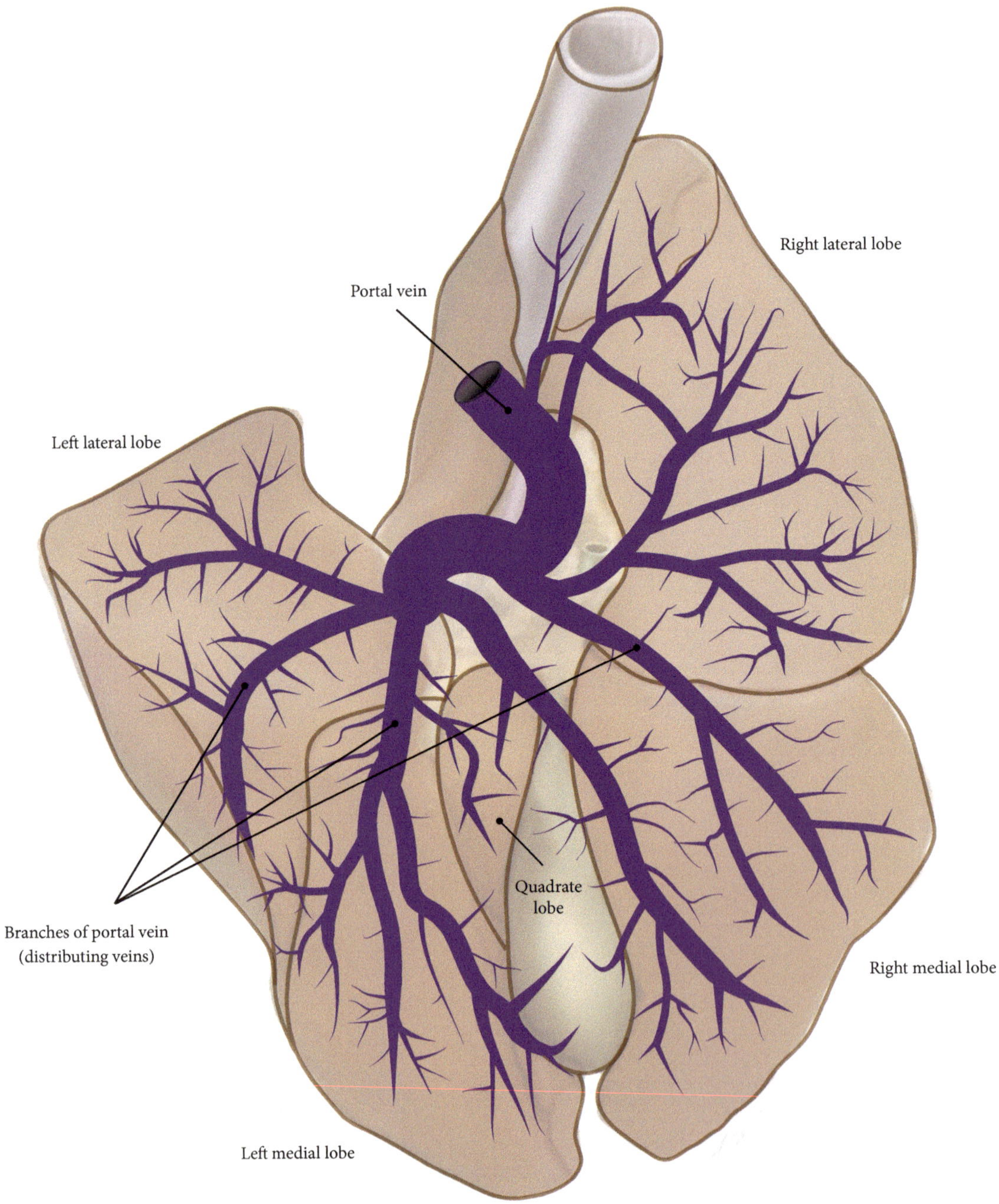

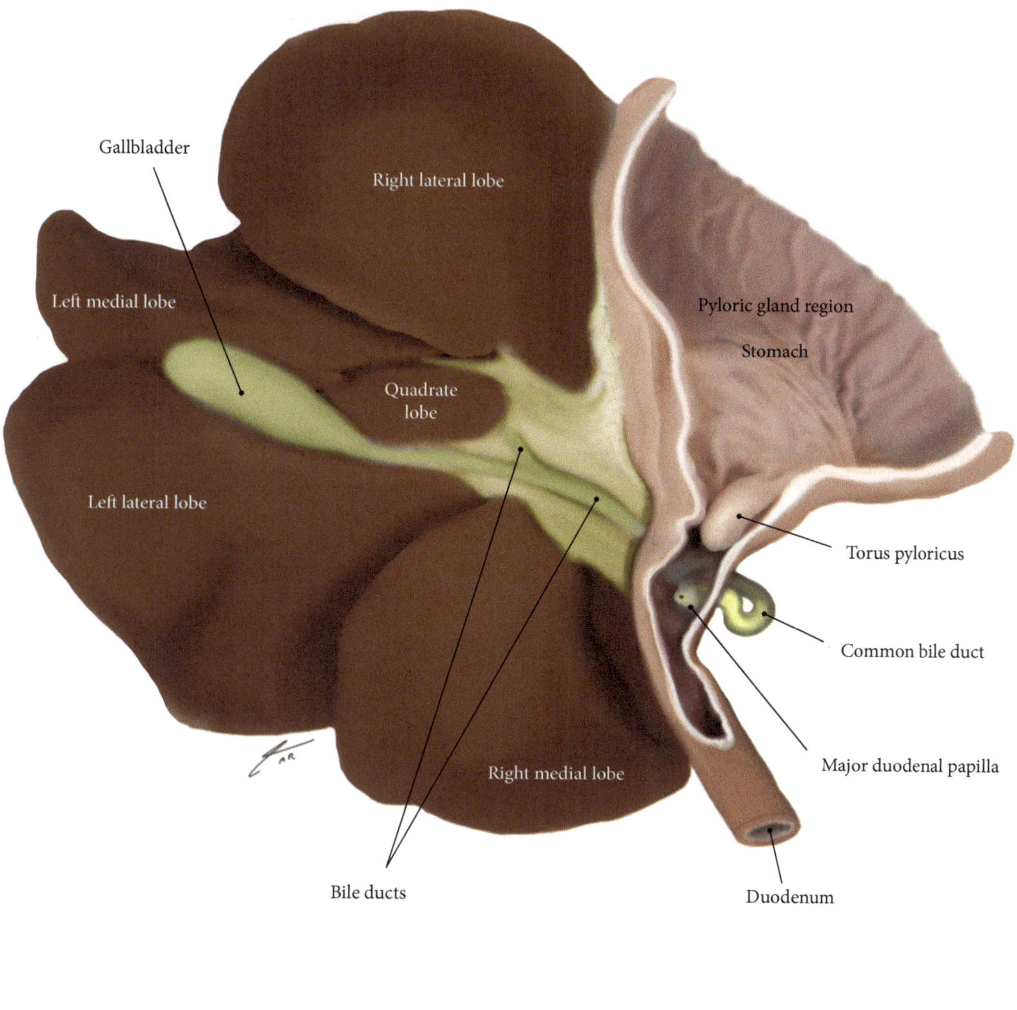
Gallbladder
Right lateral lobe
Left medial lobe
Pyloric gland region
Stomach
Quadrate lobe
Left lateral lobe
Torus pyloricus
Common bile duct
Major duodenal papilla
Right medial lobe
Bile ducts
Duodenum

Pancreas – ventral aspect

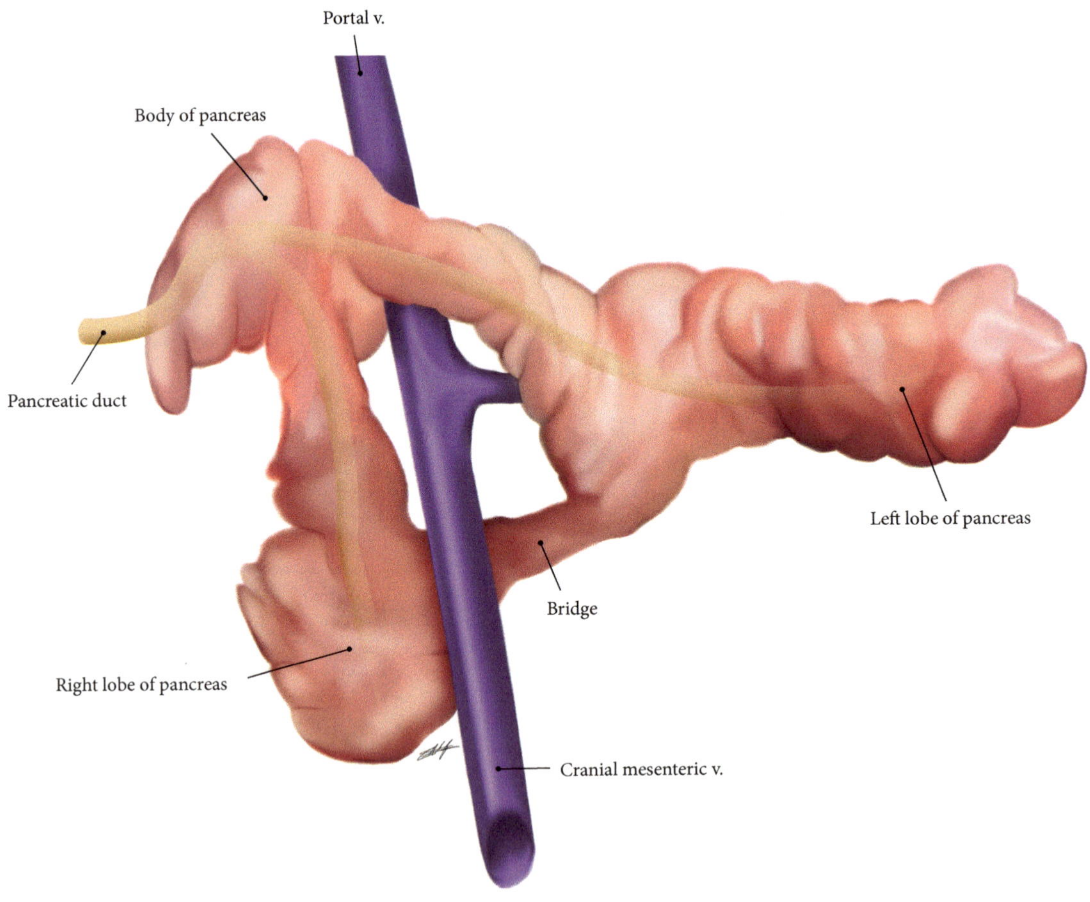

Schematic of the pancreatic blood supply and ductal system

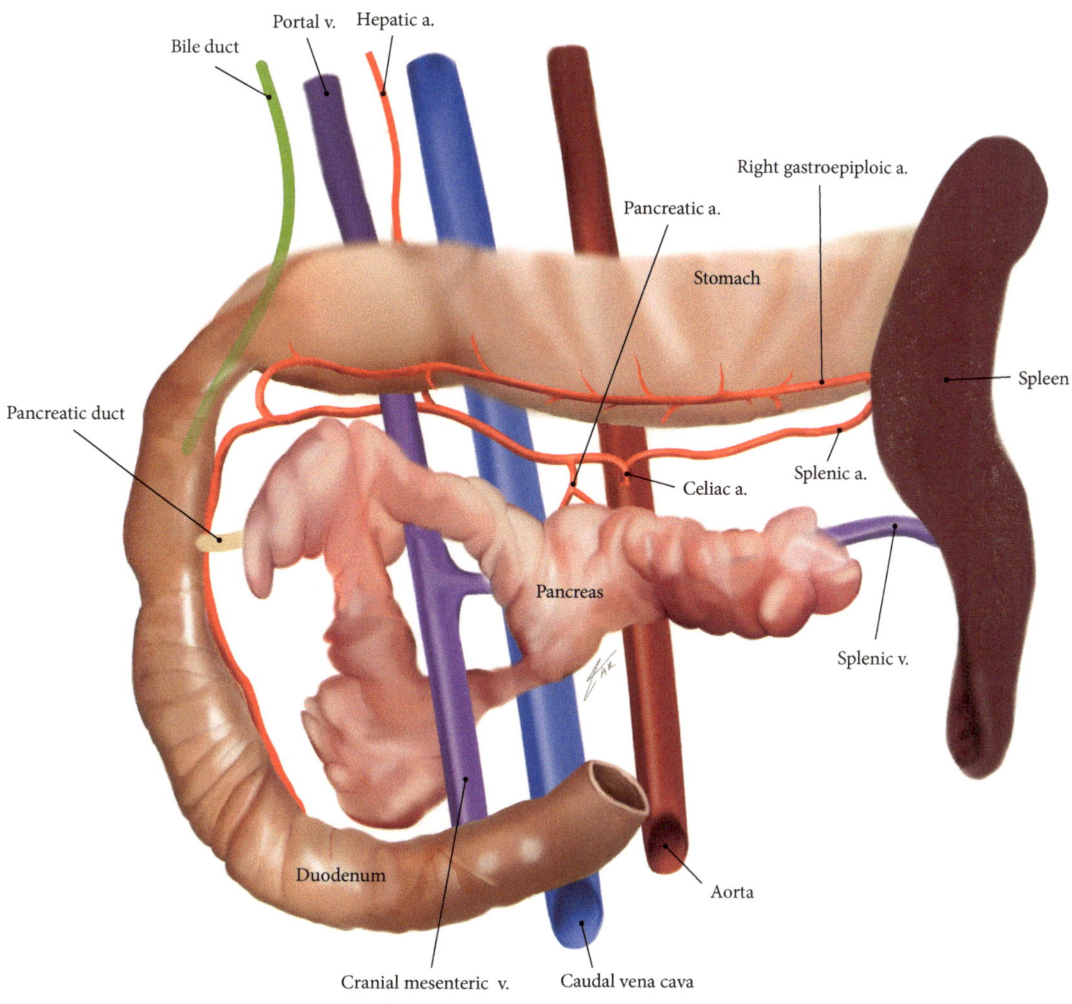

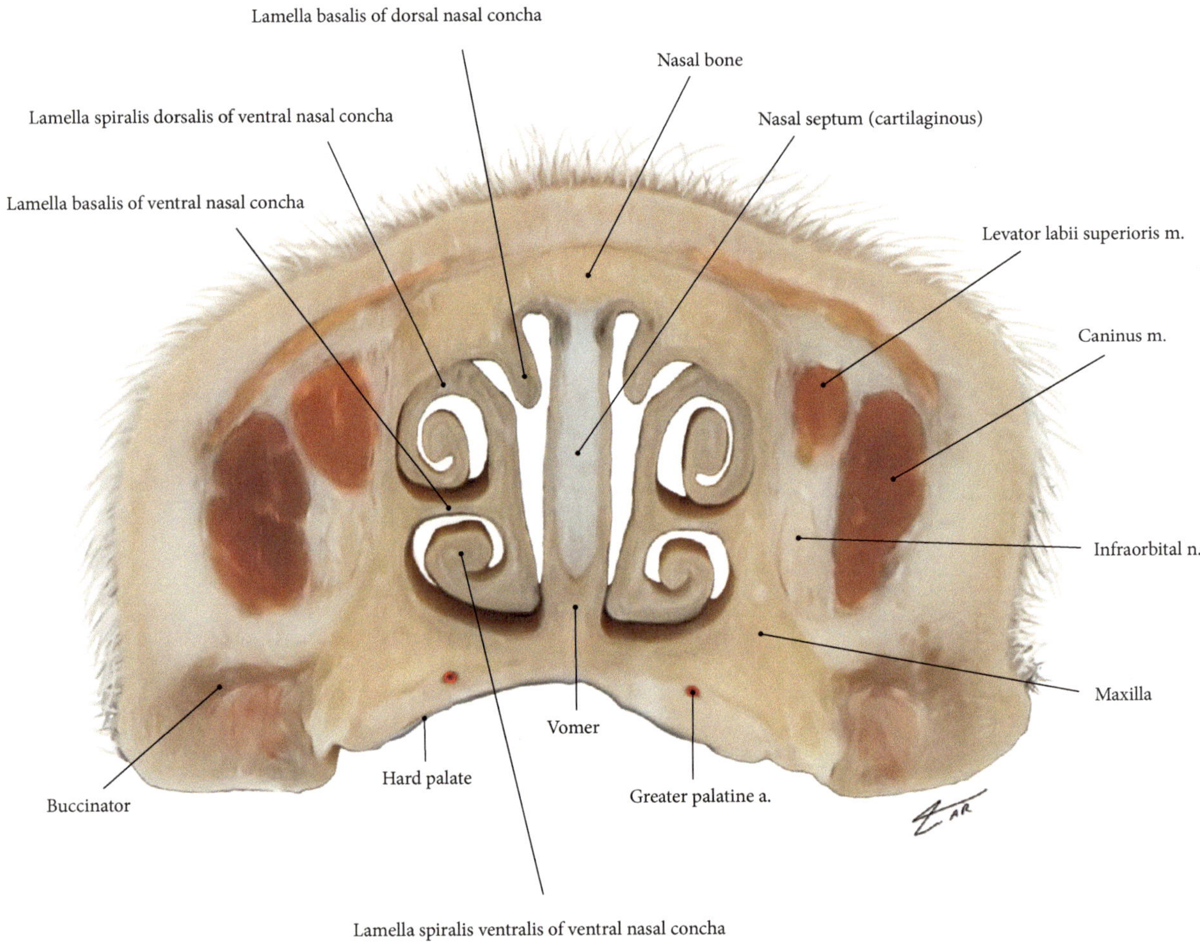
Lamella basalis of dorsal nasal concha
Nasal bone
Lamella spiralis dorsalis of ventral nasal concha
Nasal septum (cartilaginous)
Lamella basalis of ventral nasal concha
Levator labii superioris m.
Caninus m.
Infraorbital n.
Maxilla
Vomer
Hard palate
Greater palatine a.
Buccinator
Lamella spiralis ventralis of ventral nasal concha

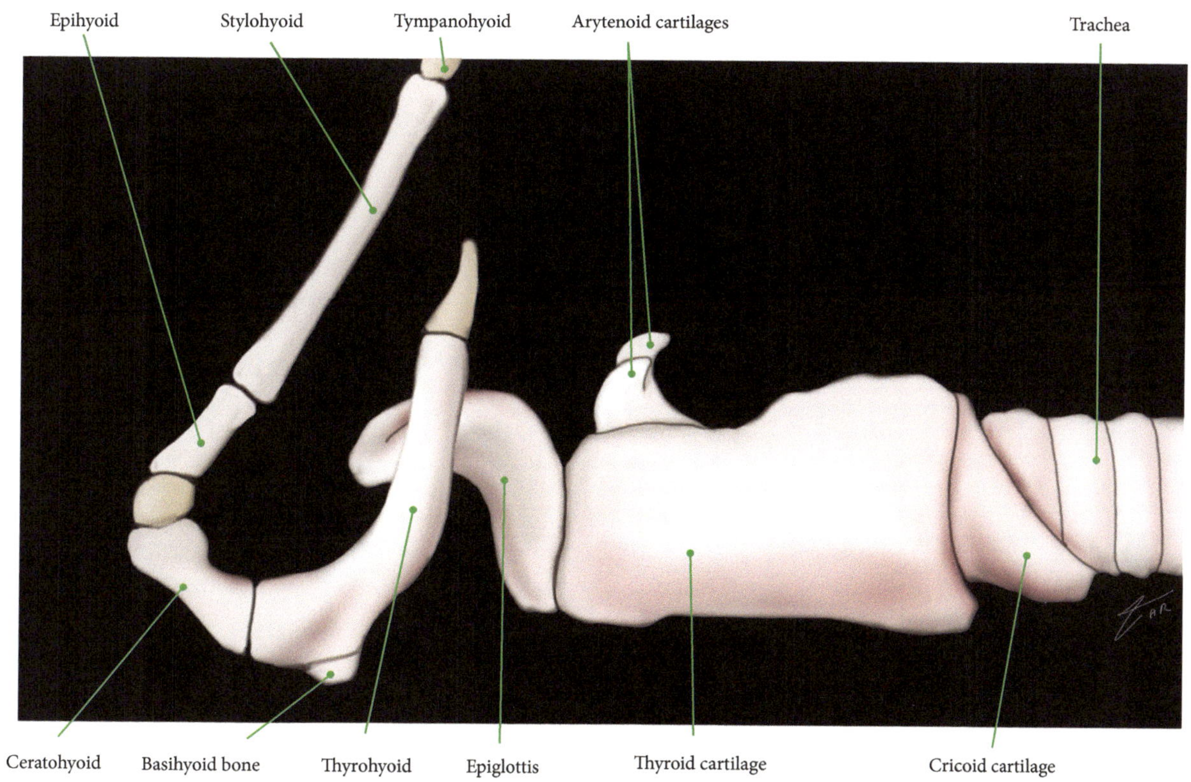
Epihyoid
Stylohyoid
Tympanohyoid
Arytenoid cartilages
Trachea
Ceratohyoid
Basihyoid bone
Thyrohyoid
Epiglottis
Thyroid cartilage
Cricoid cartilage

Larynx – median section

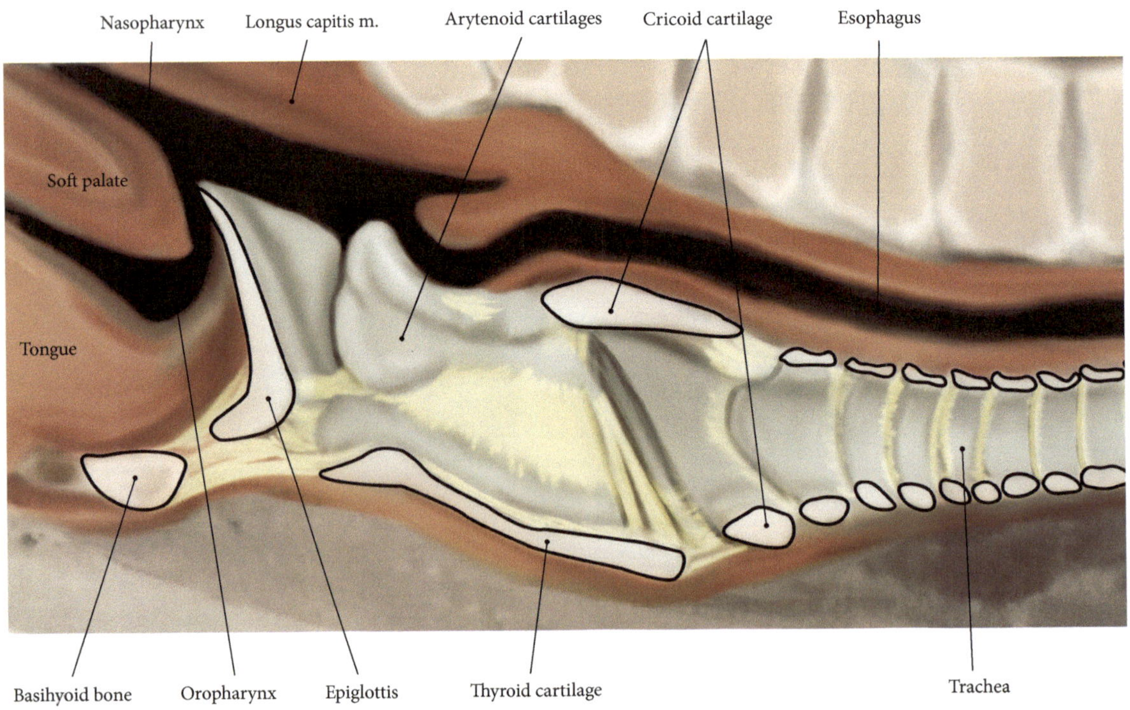

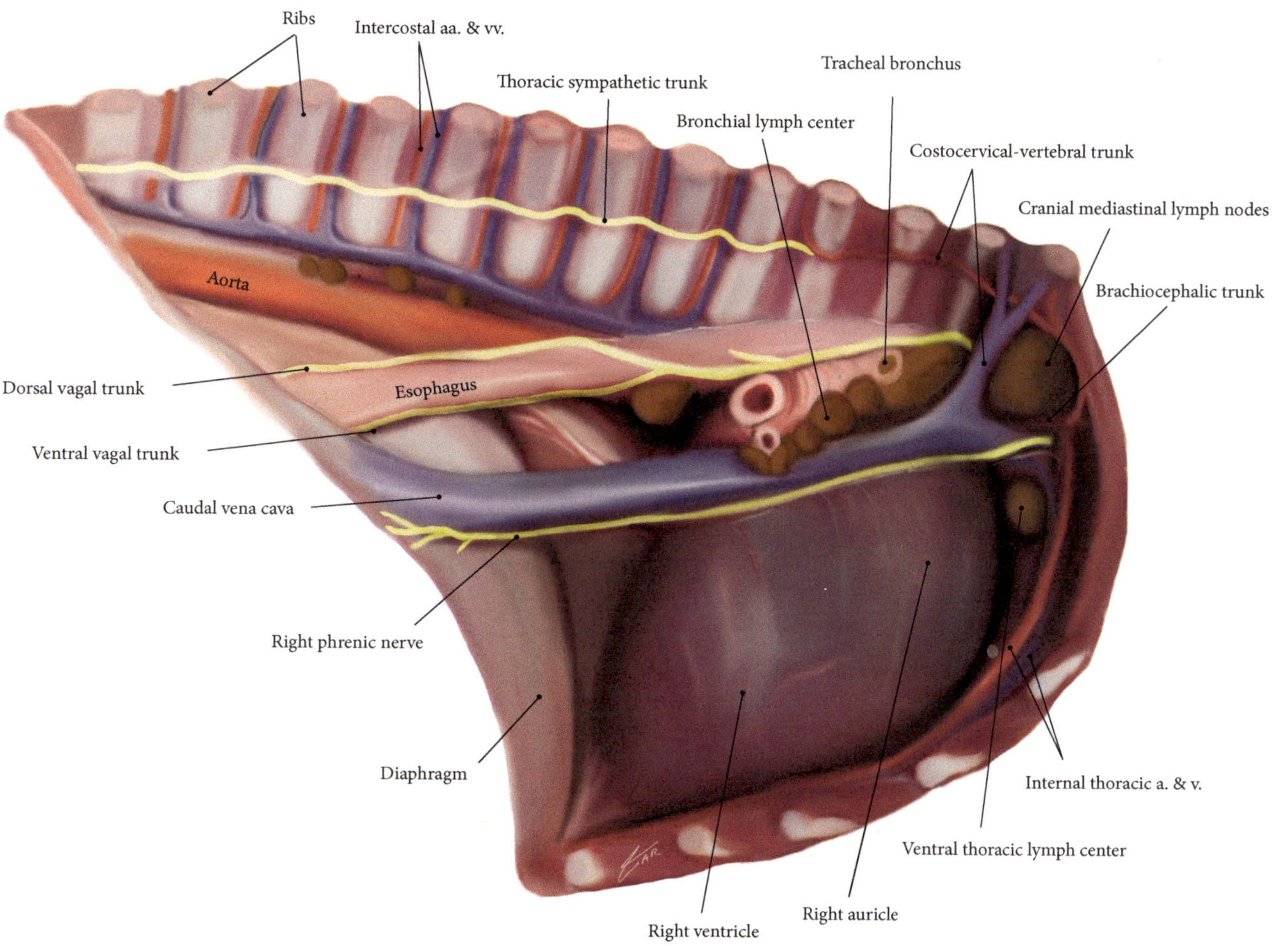

Ribs
Intercostal aa. & vv.
Thoracic sympathetic trunk
Tracheal bronchus
Bronchial lymph center
Costocervical-vertebral trunk
Cranial mediastinal lymph nodes
Brachiocephalic trunk
Aorta
Dorsal vagal trunk
Esophagus
Ventral vagal trunk
Caudal vena cava
Right phrenic nerve
Diaphragm
Internal thoracic a. & v.
Ventral thoracic lymph center
Right auricle
Right ventricle

Left lateral thorax

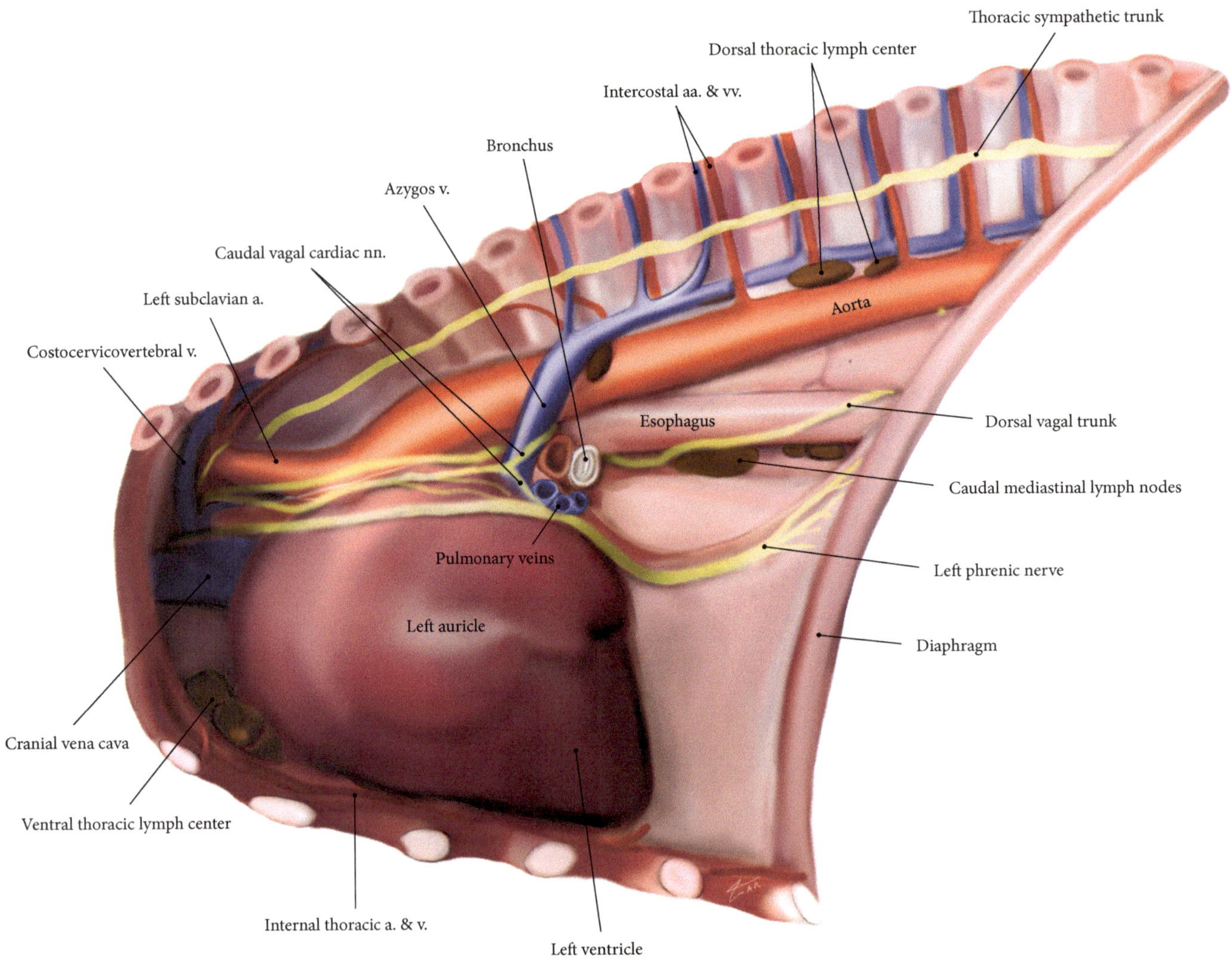

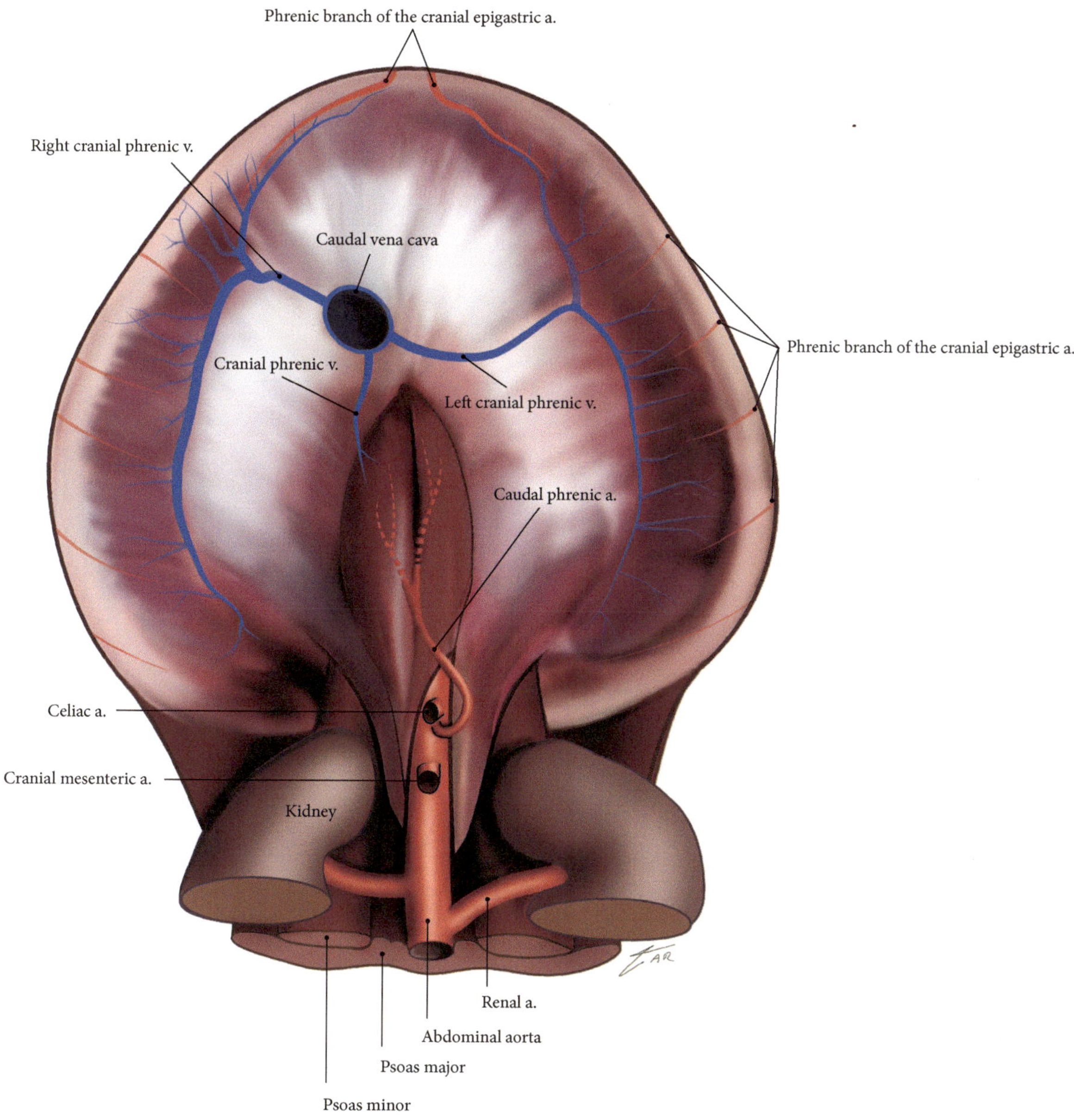
Phrenic branch of the cranial epigastric a.
Right cranial phrenic v.
Caudal vena cava
Phrenic branch of the cranial epigastric a.
Cranial phrenic v.
Left cranial phrenic v.
Caudal phrenic a.
Celiac a.
Cranial mesenteric a.
Kidney
Renal a.
Abdominal aorta
Psoas major
Psoas minor

Left lung – costal surface

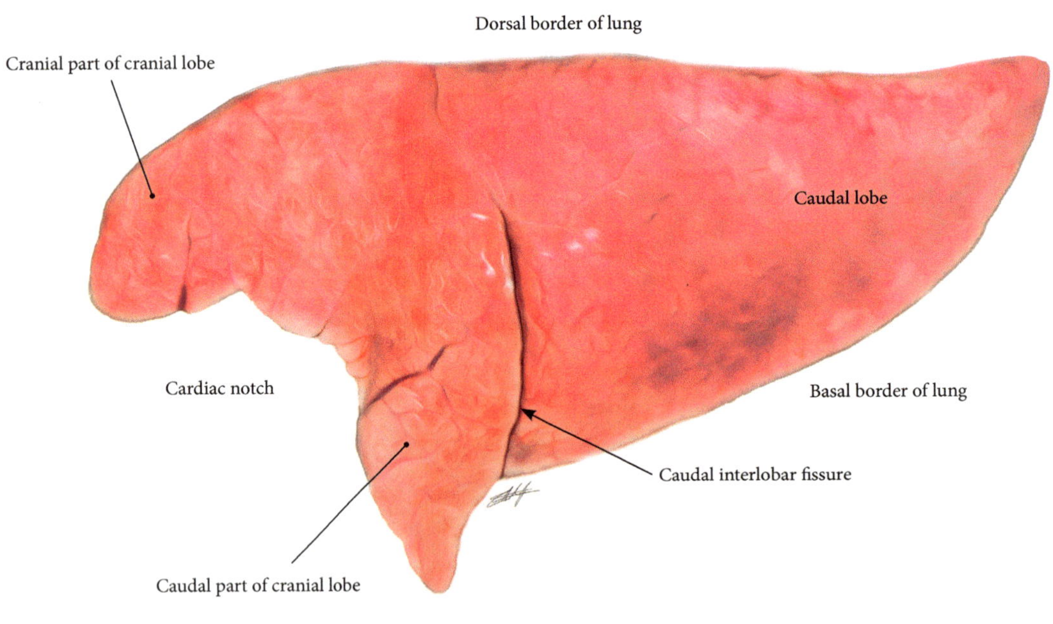

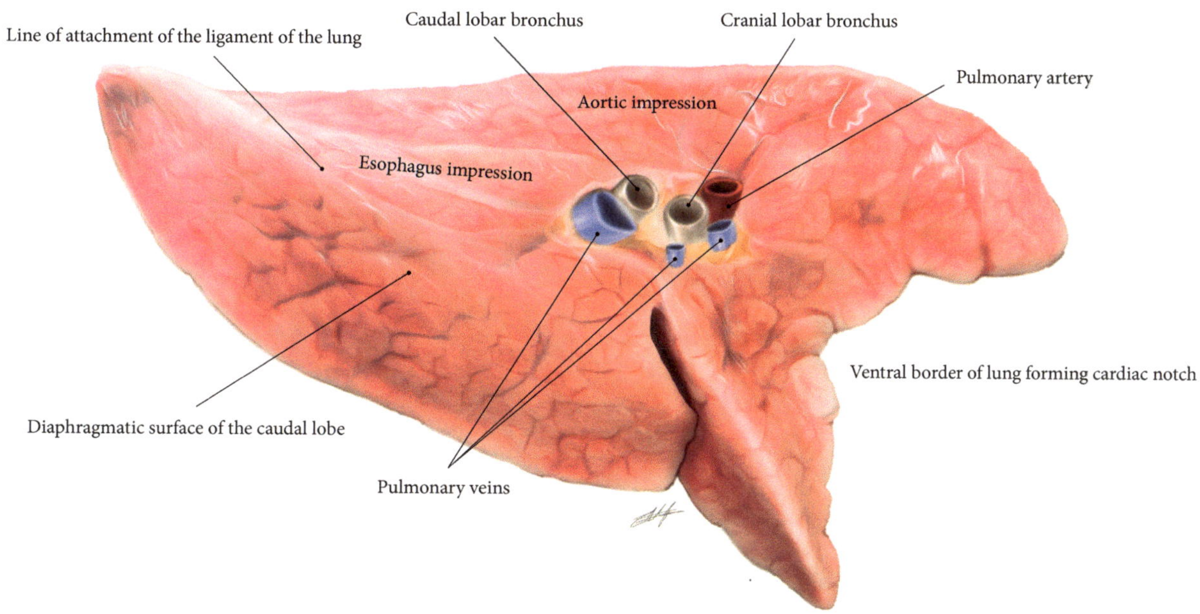

Line of attachment of the ligament of the lung
Caudal lobar bronchus
Cranial lobar bronchus
Pulmonary artery
Aortic impression
Esophagus impression
Ventral border of lung forming cardiac notch
Diaphragmatic surface of the caudal lobe
Pulmonary veins

Right lung – costal surface

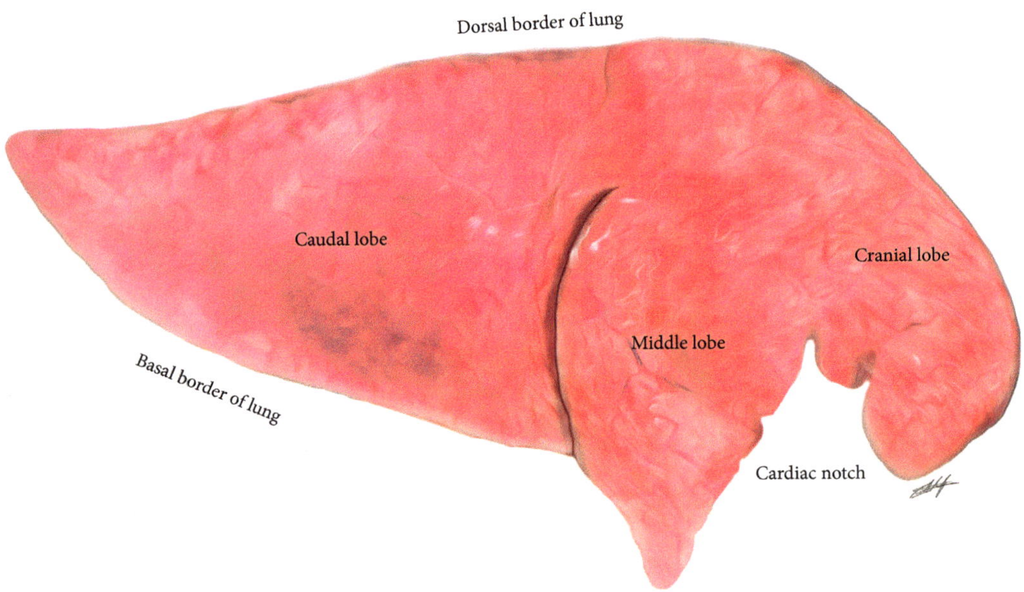

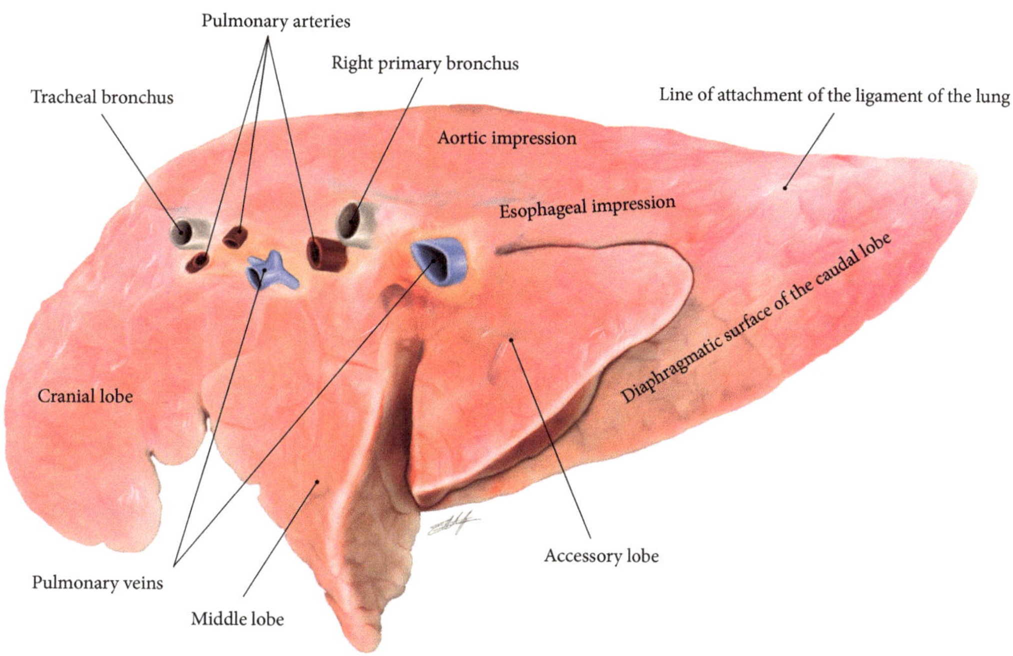
Pulmonary arteries
Right primary bronchus
Tracheal bronchus
Line of attachment of the ligament of the lung
Aortic impression
Esophageal impression
Diaphragmatic surface of the caudal lobe
Cranial lobe
Accessory lobe
Pulmonary veins
Middle lobe

Trachea and bronchial tree – dorsal aspect

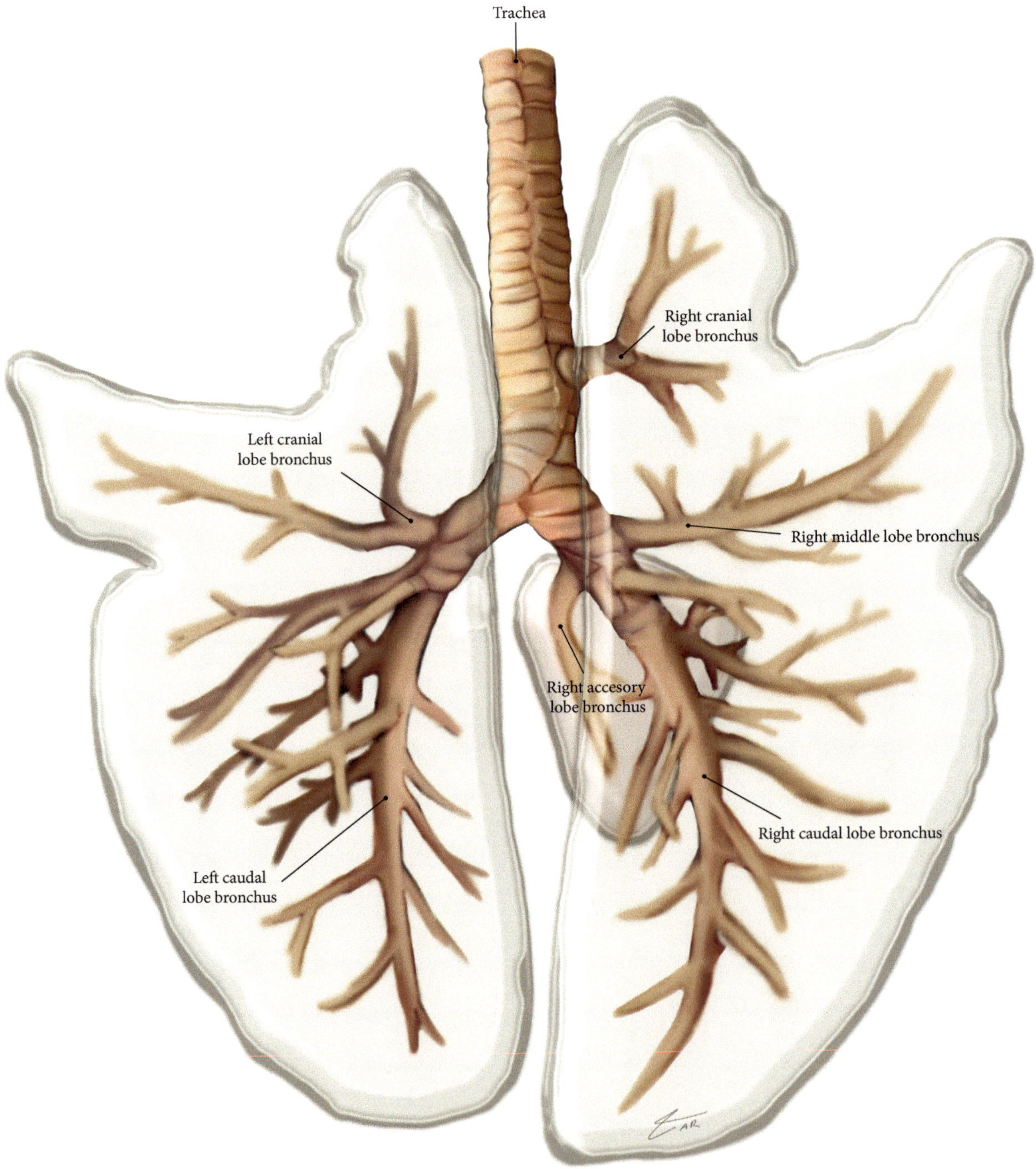

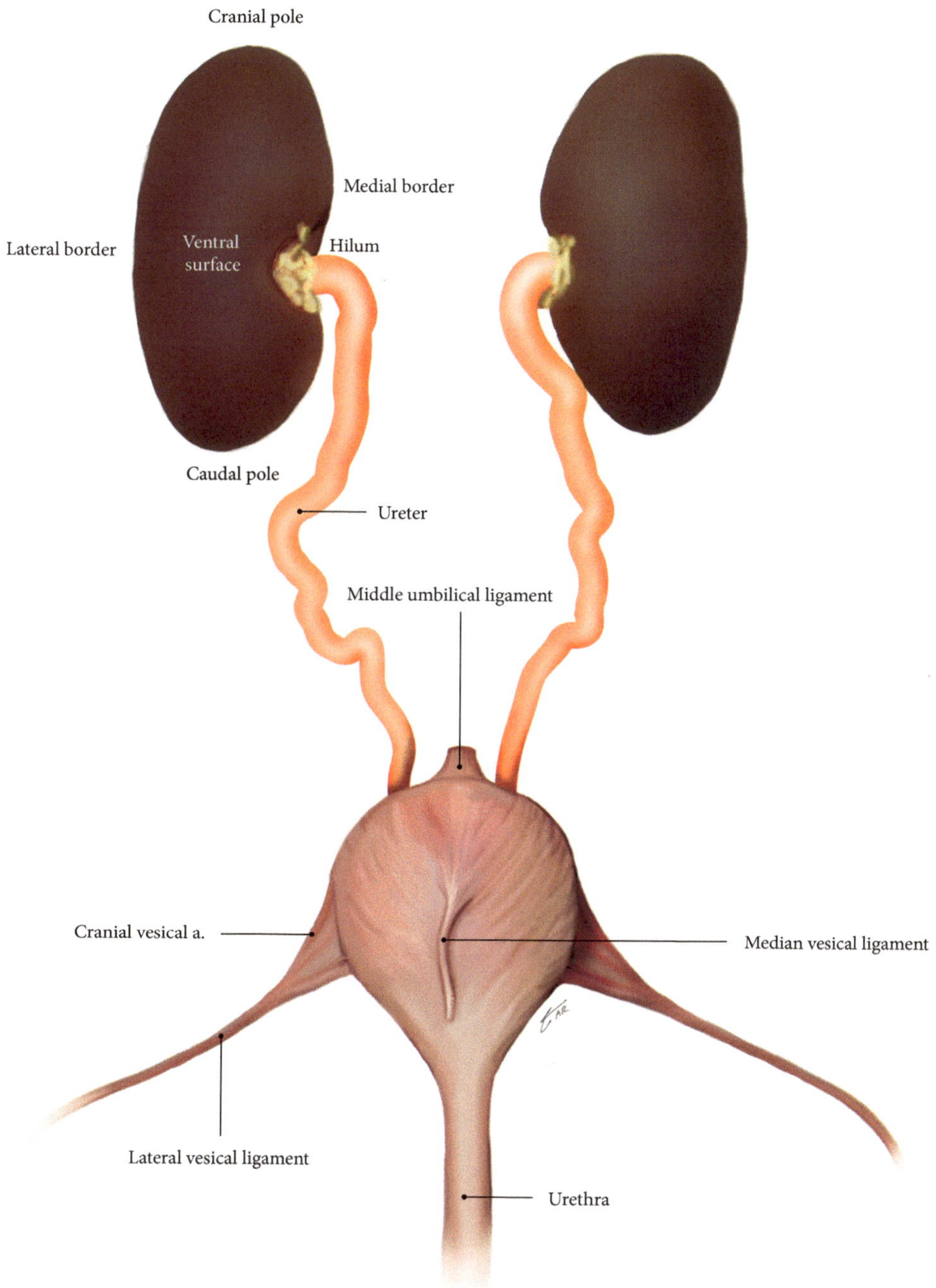
Cranial pole
Medial border
Lateral border
Ventral
surface
Hilum
Caudal pole
Ureter
Middle umbilical ligament
Cranial vesical a.
Median vesical ligament
Lateral vesical ligament
Urethra

Kidney – right and left

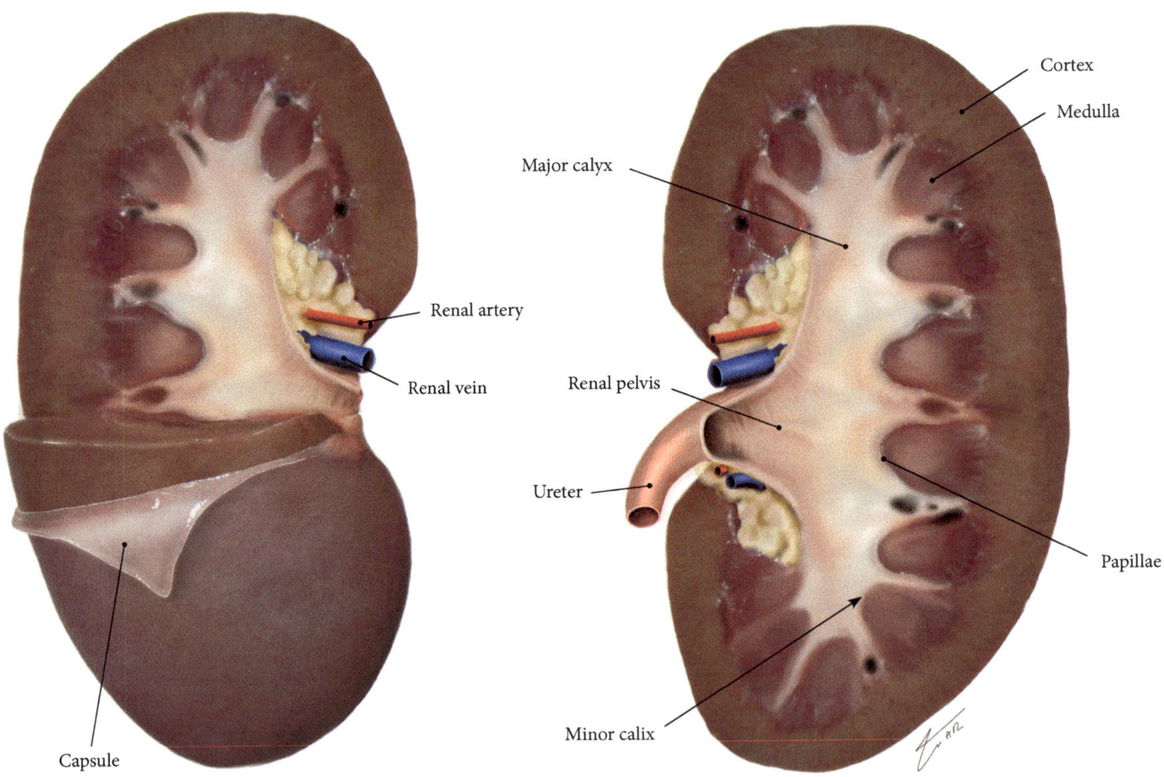

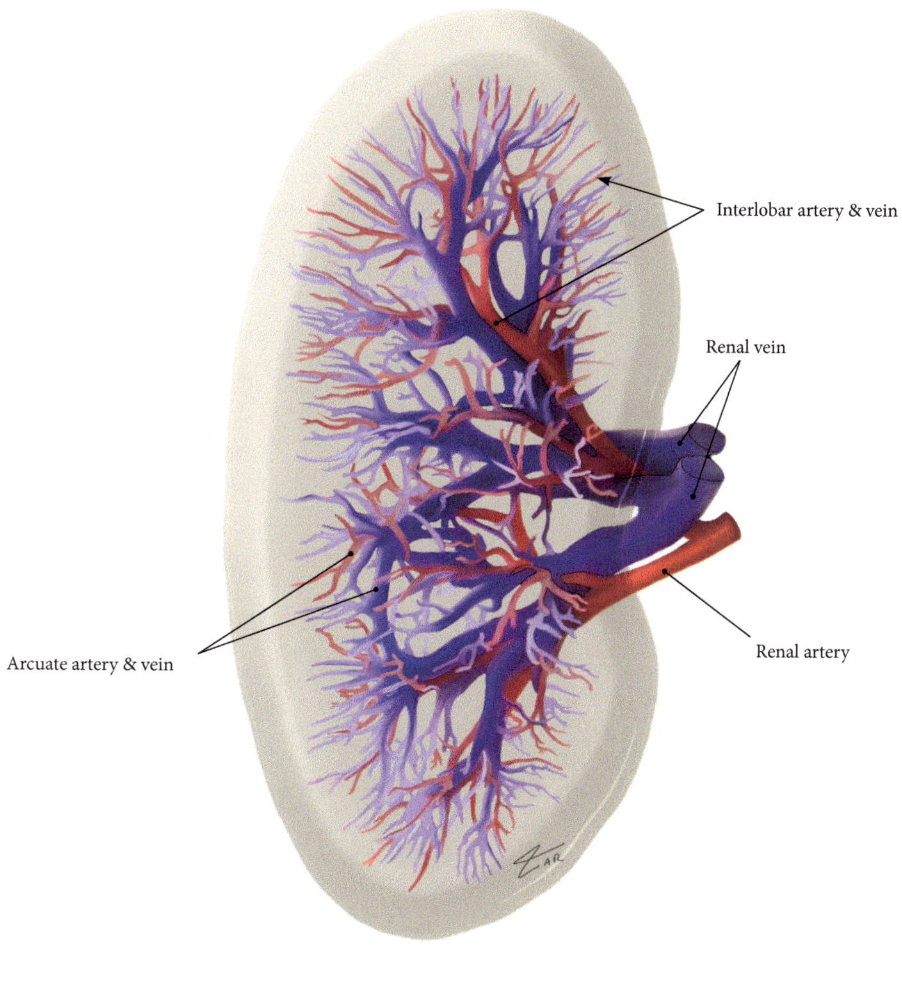
Interlobar artery & vein
Renal vein
Renal artery
Arcuate artery & vein

Kidney – cast of the renal pelvis

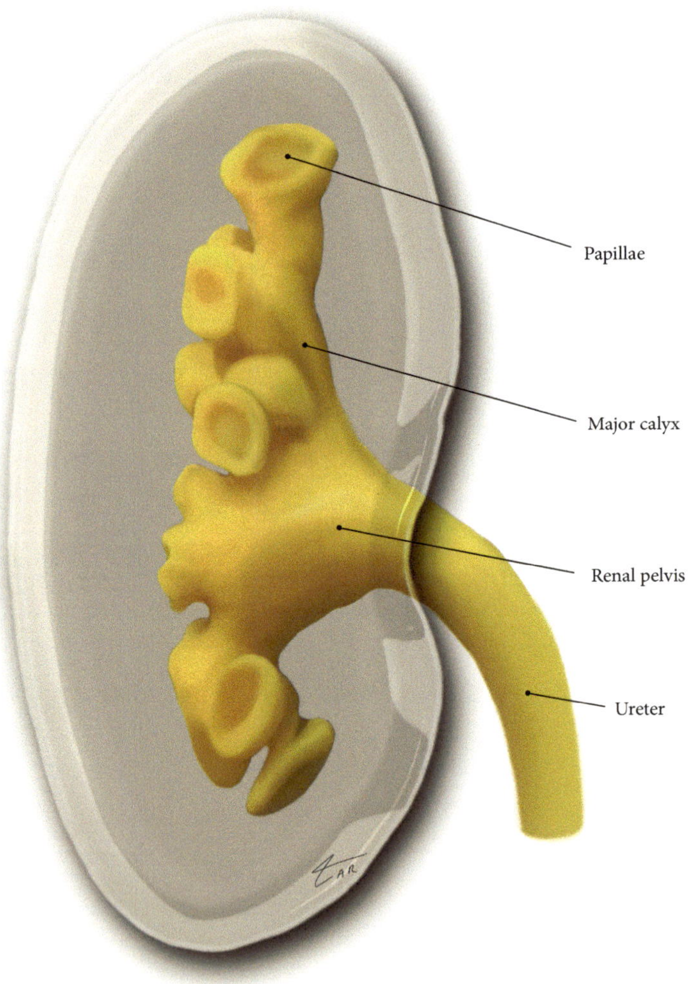

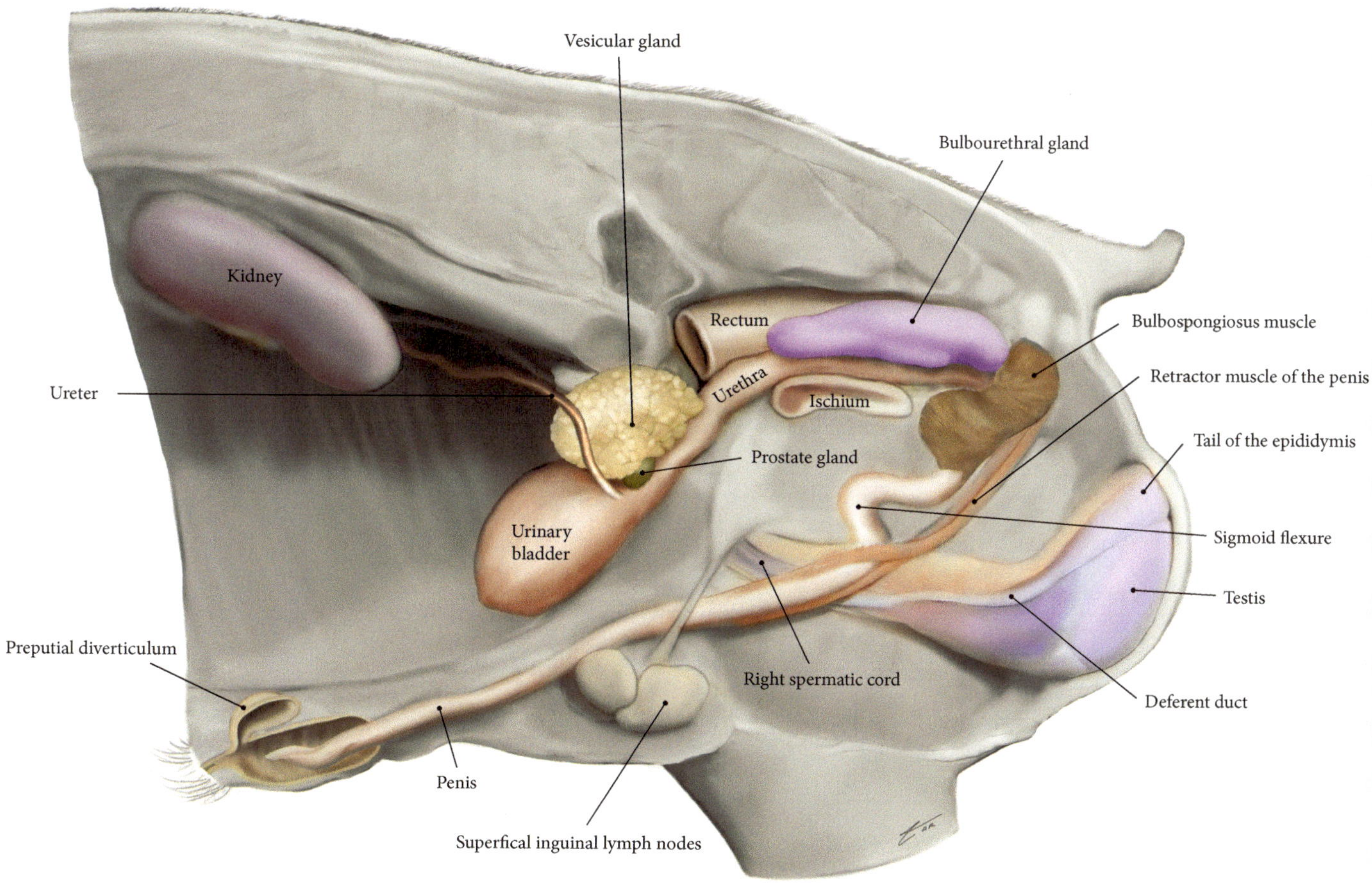
Vesicular gland
Bulbourethral gland
Kidney
Rectum
Bulbospongiosus muscle
Urethra
Retractor muscle of the penis
Ureter
Ischium
Tail of the epididymis
Prostate gland
Urinary bladder
Sigmoid flexure
Testis
Preputial diverticulum
Right spermatic cord
Deferent duct
Penis
Superfical inguinal lymph nodes

Sagittal section – preputial diverticulum and prepuce

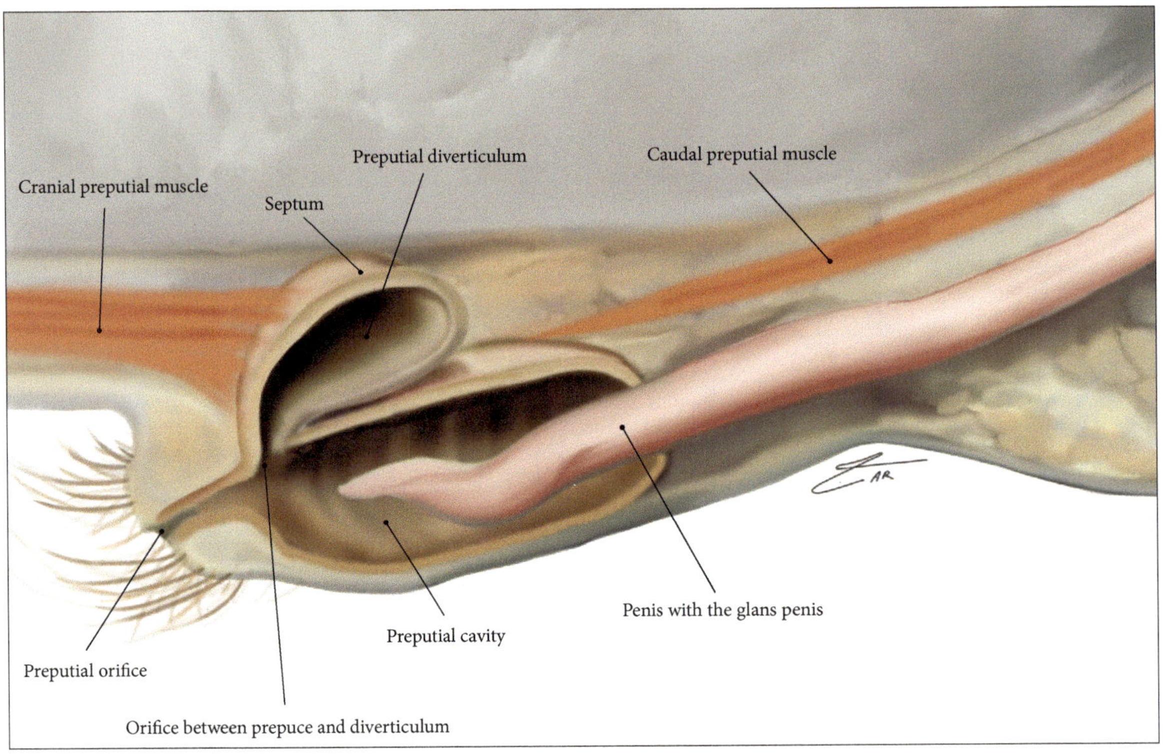

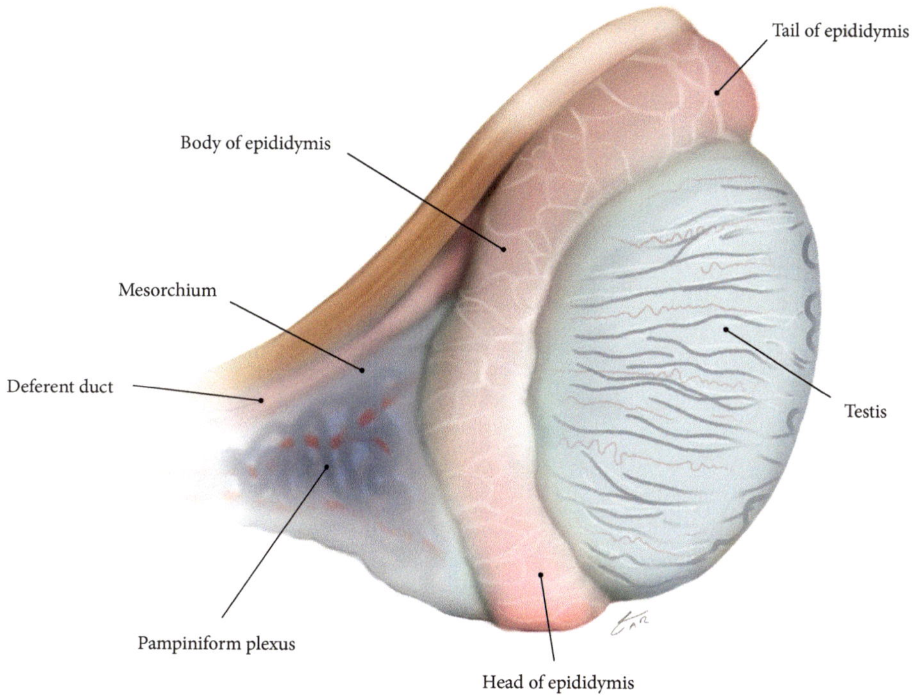
Tail of epididymis
Body of epididymis
Mesorchium
Deferent duct
Testis
Pampiniform plexus
Head of epididymis

Testicle – medial aspect

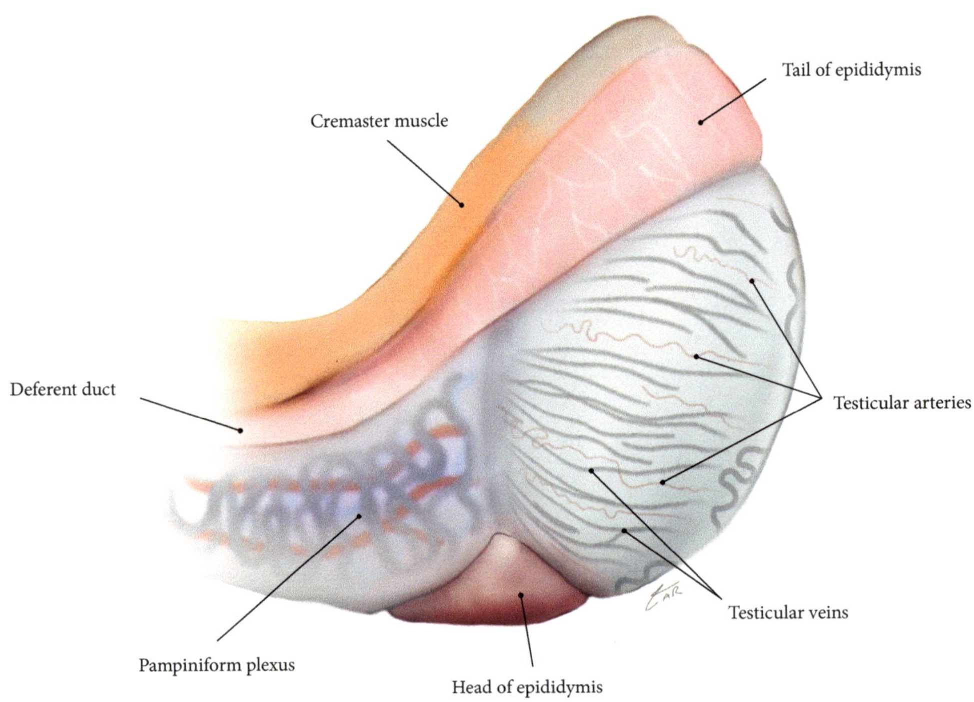

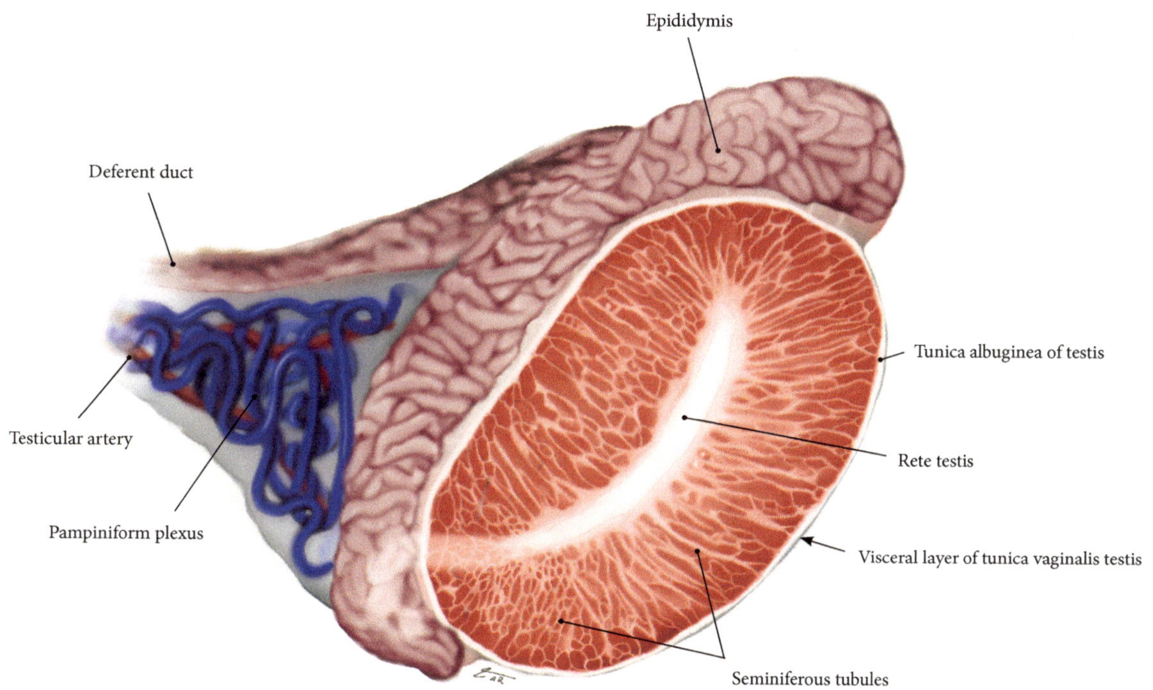
Epididymis
Deferent duct
Tunica albuginea of testis
Testicular artery
Rete testis
Pampiniform plexus
Visceral layer of tunica vaginalis testis
Seminiferous tubules

Arteries to the genital tract of the sow

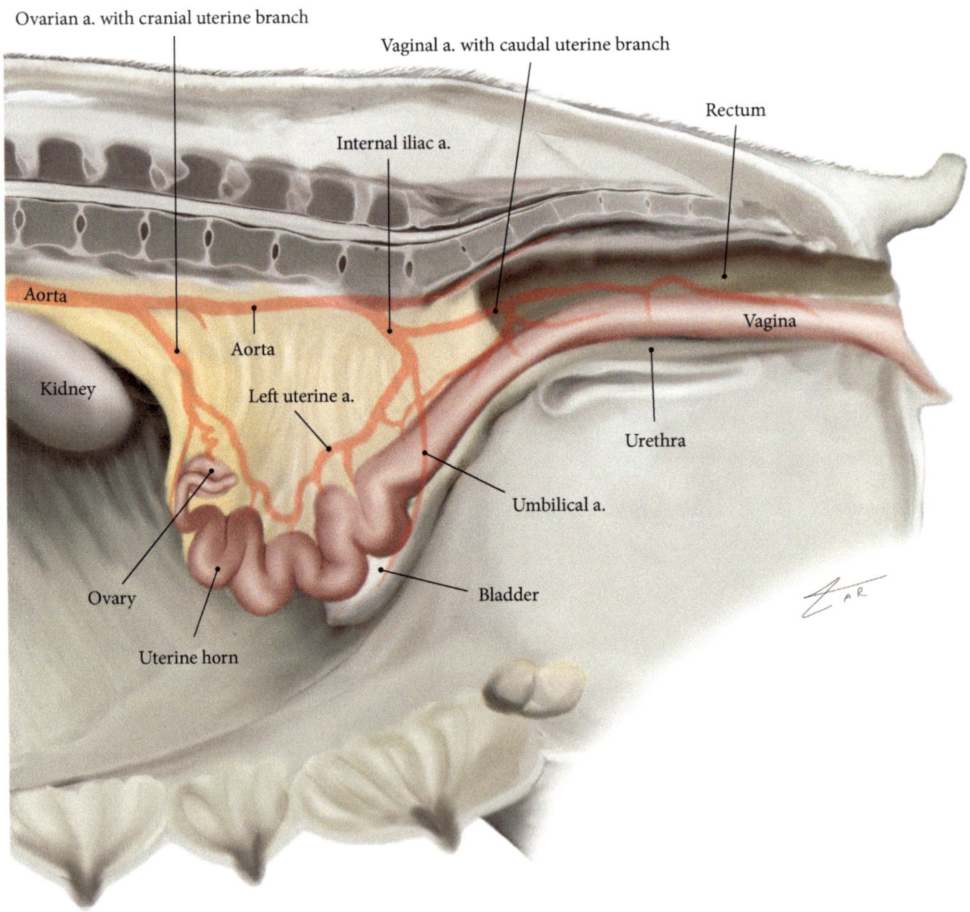

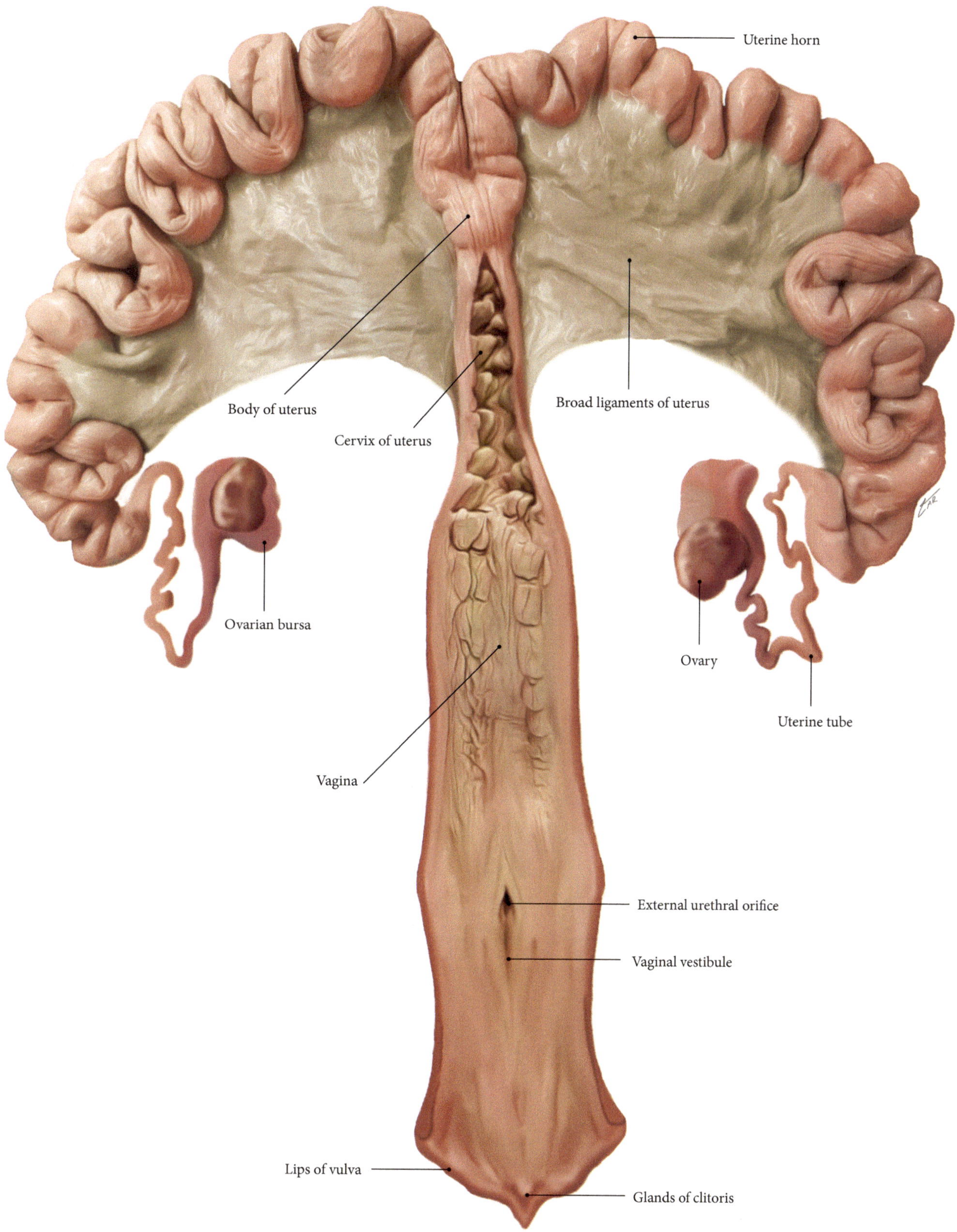
Uterine horn
Body of uterus
Cervix of uterus
Broad ligaments of uterus
Ovarian bursa
Ovary
Uterine tube
Vagina
External urethral orifice
Vaginal vestibule
Lips of vulva
Glands of clitoris

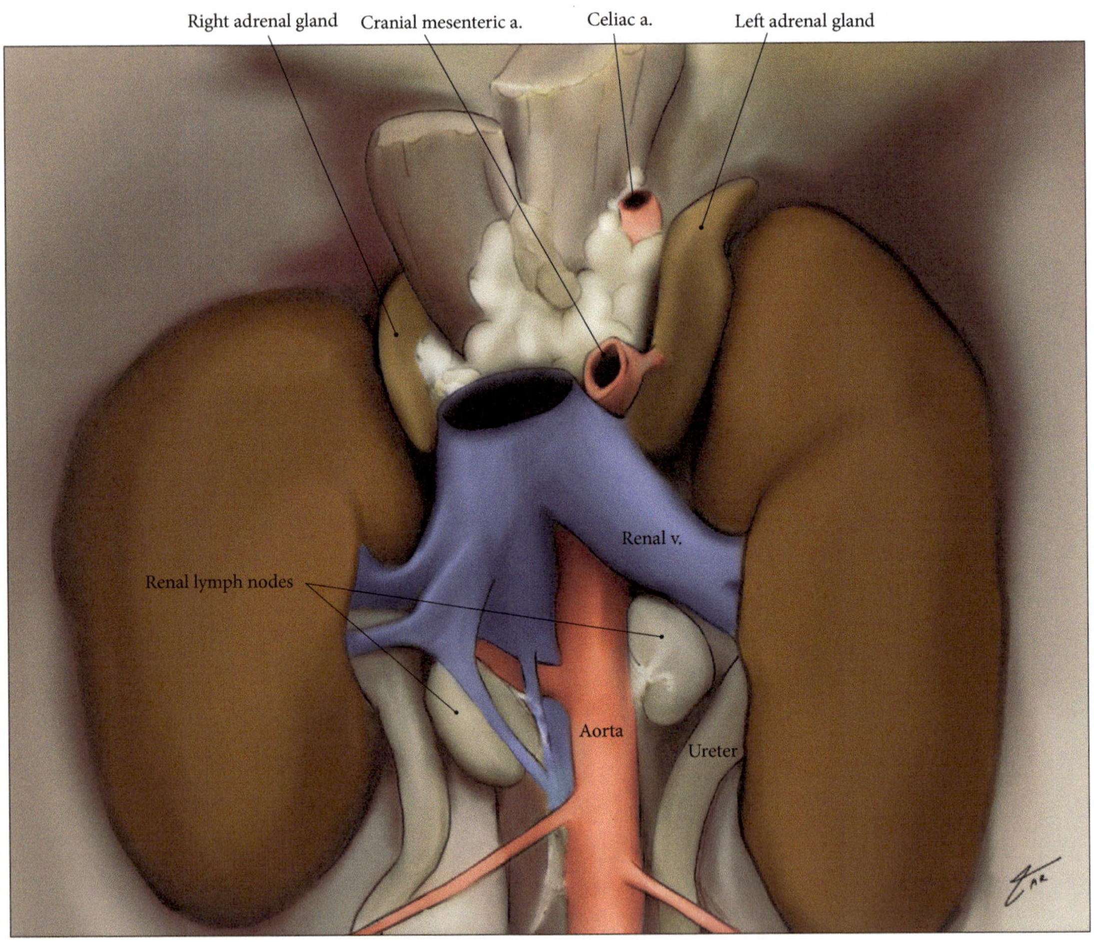
Right adrenal gland
Cranial mesenteric a.
Celiac a.
Left adrenal gland
Renal v.
Renal lymph nodes
Aorta
Ureter

Dorsal view

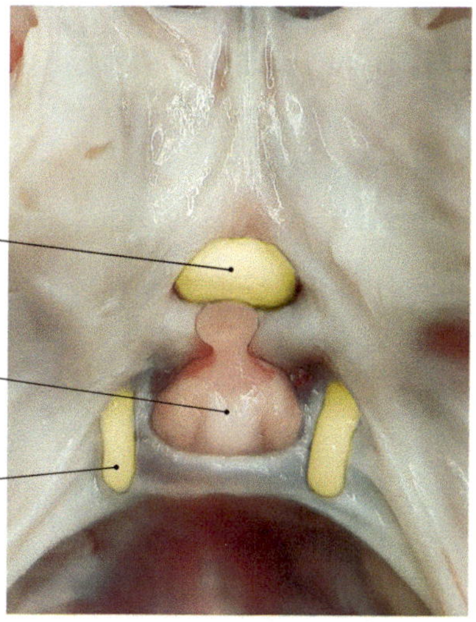

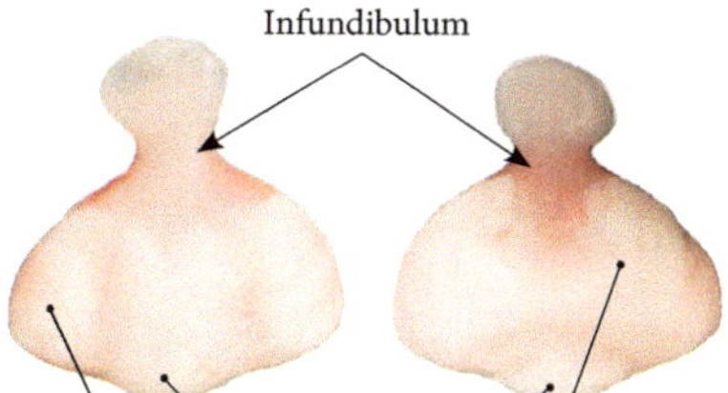

Midsagittal section

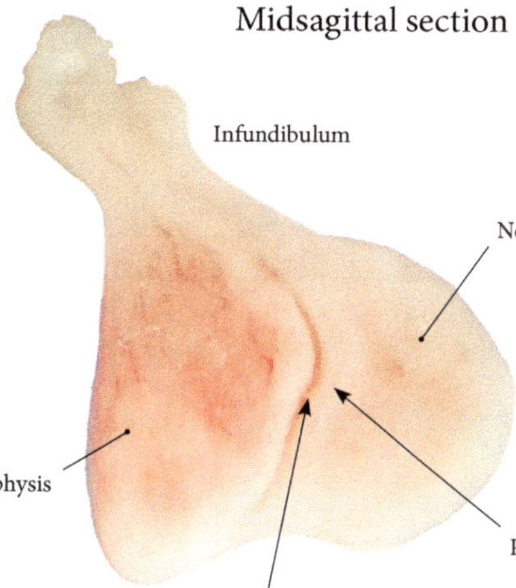

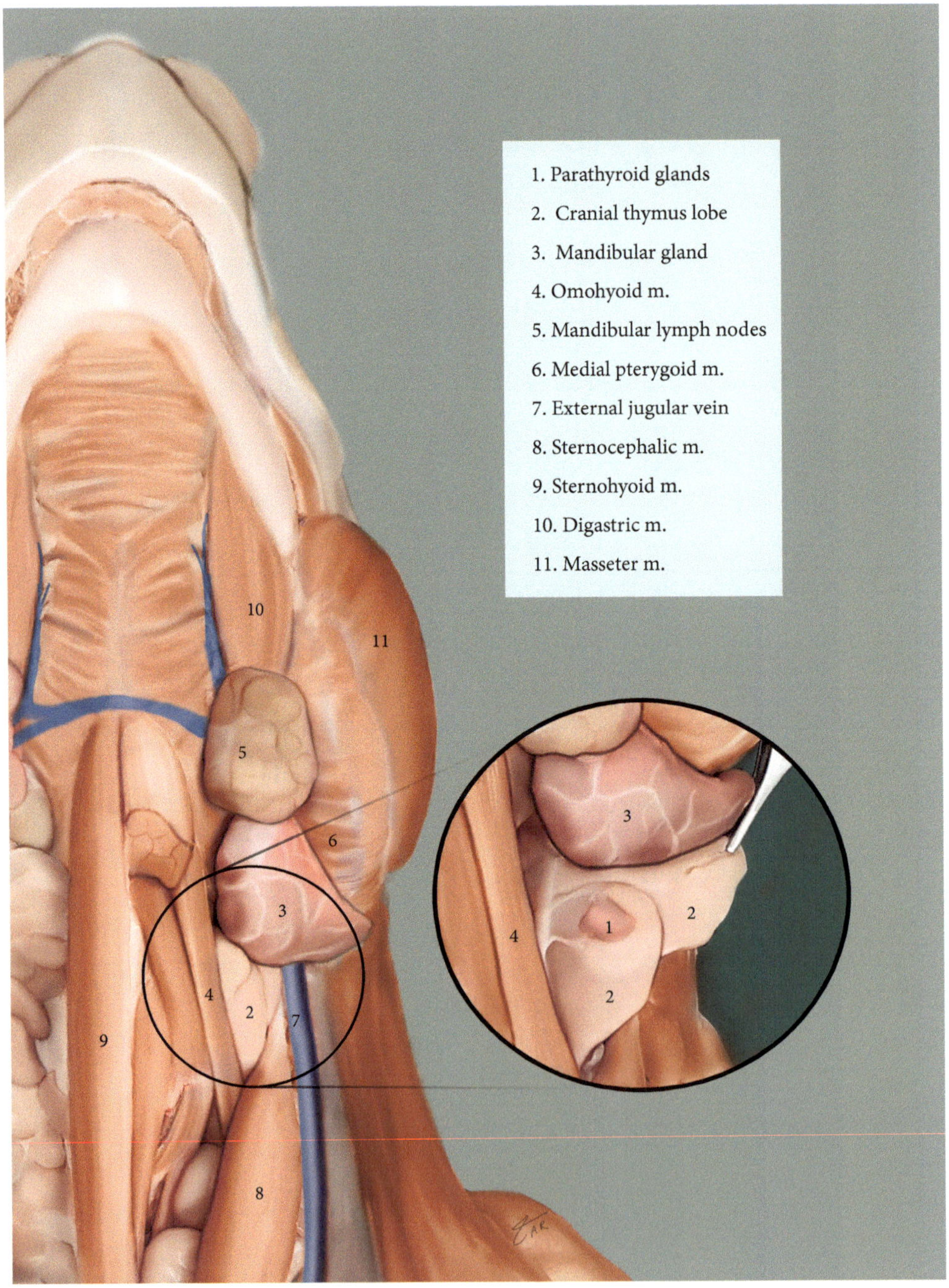
1. Parathyroid glands
2. Cranial thymus lobe
3. Mandibular gland
4. Omohyoid m.
5. Mandibular lymph nodes
6. Medial pterygoid m.
7. External jugular vein
8. Sternocephalic m.
9. Sternohyoid m.
10. Digastric m.
11. Masseter m.

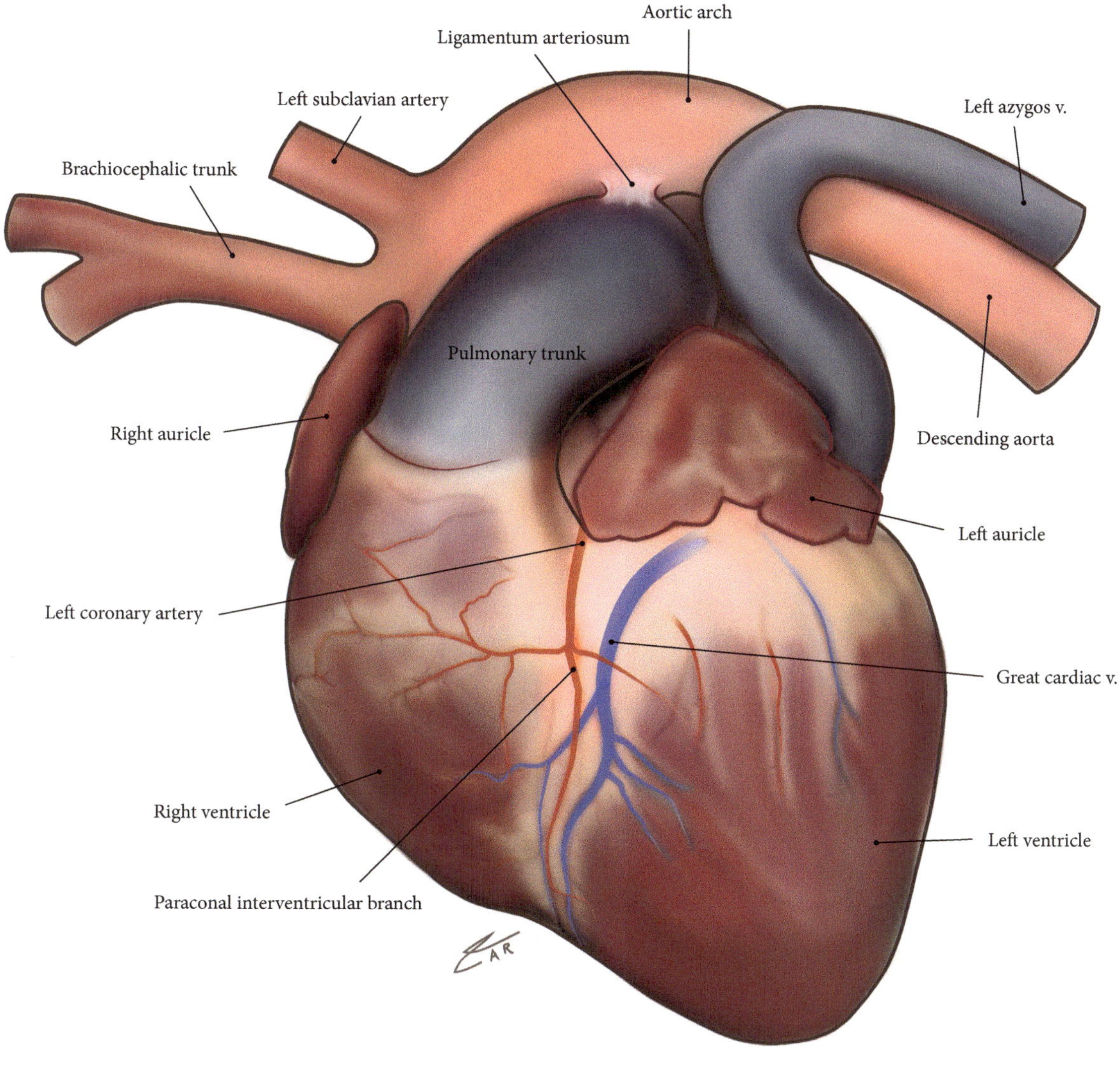
Aortic arch
Ligamentum arteriosum
Left subclavian artery
Left azygos v.
Brachiocephalic trunk
Pulmonary trunk
Right auricle
Descending aorta
Left auricle
Left coronary artery
Great cardiac v.
Right ventricle
Left ventricle
Paraconal interventricular branch

Heart of pig – atrial (right) surface

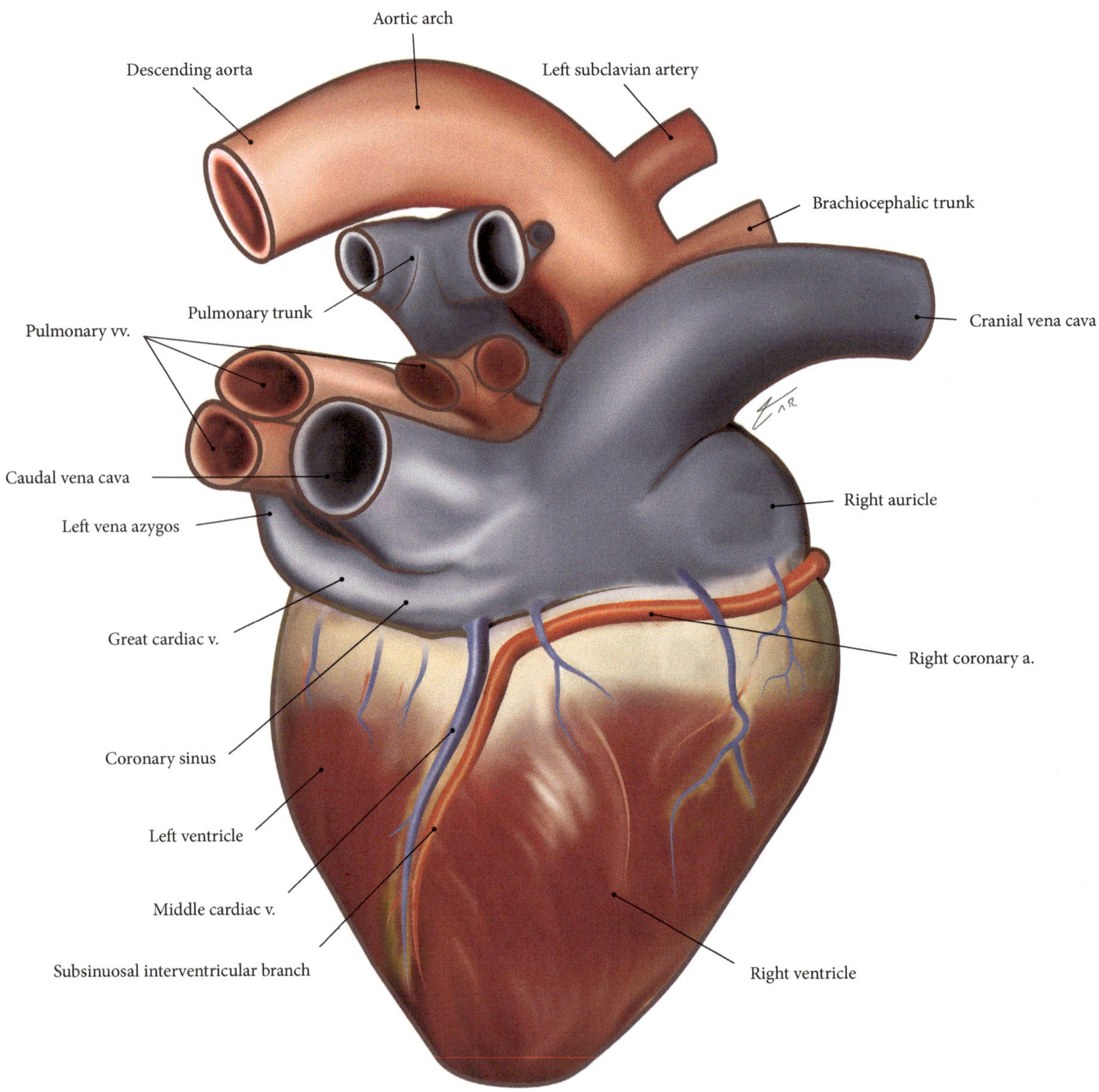

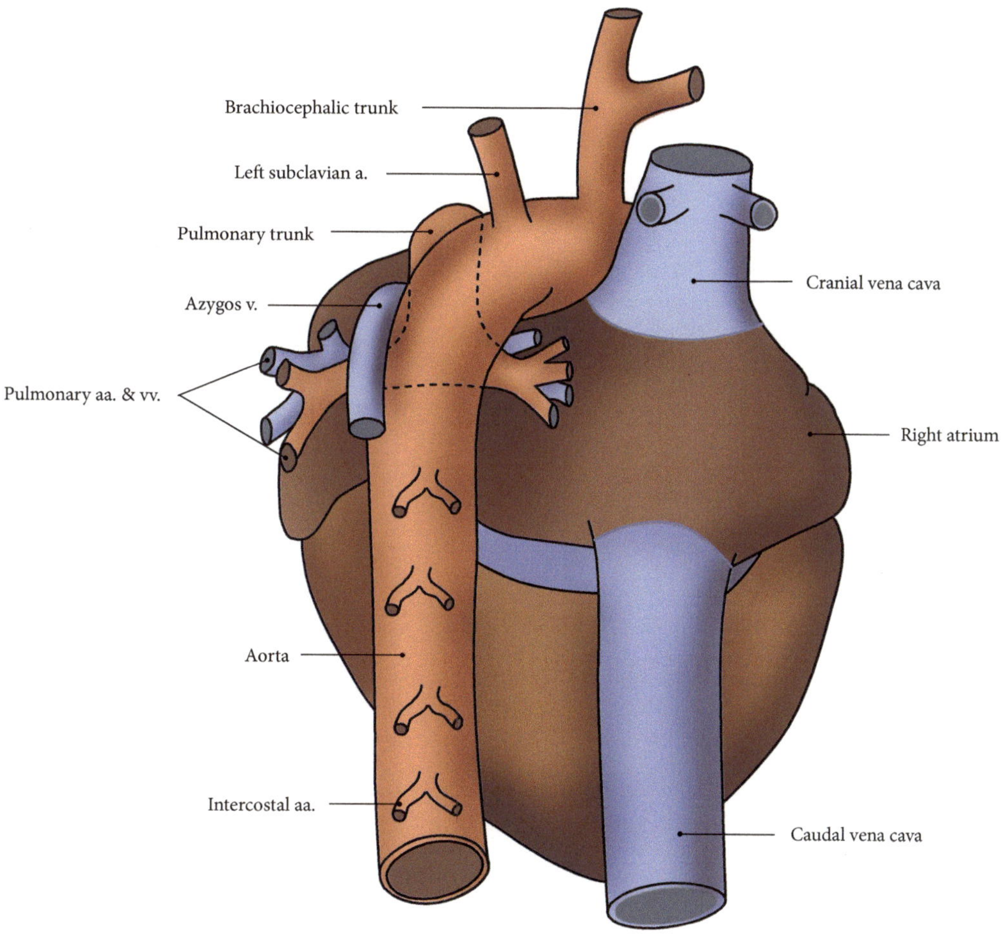
Brachiocephalic trunk
Left subclavian a.
Pulmonary trunk
Azygos v.
Pulmonary aa. & vv.
Aorta
Intercostal aa.
Cranial vena cava
Right atrium
Caudal vena cava

Dorsal view of the base of the heart – greater blood vessels and atria transected

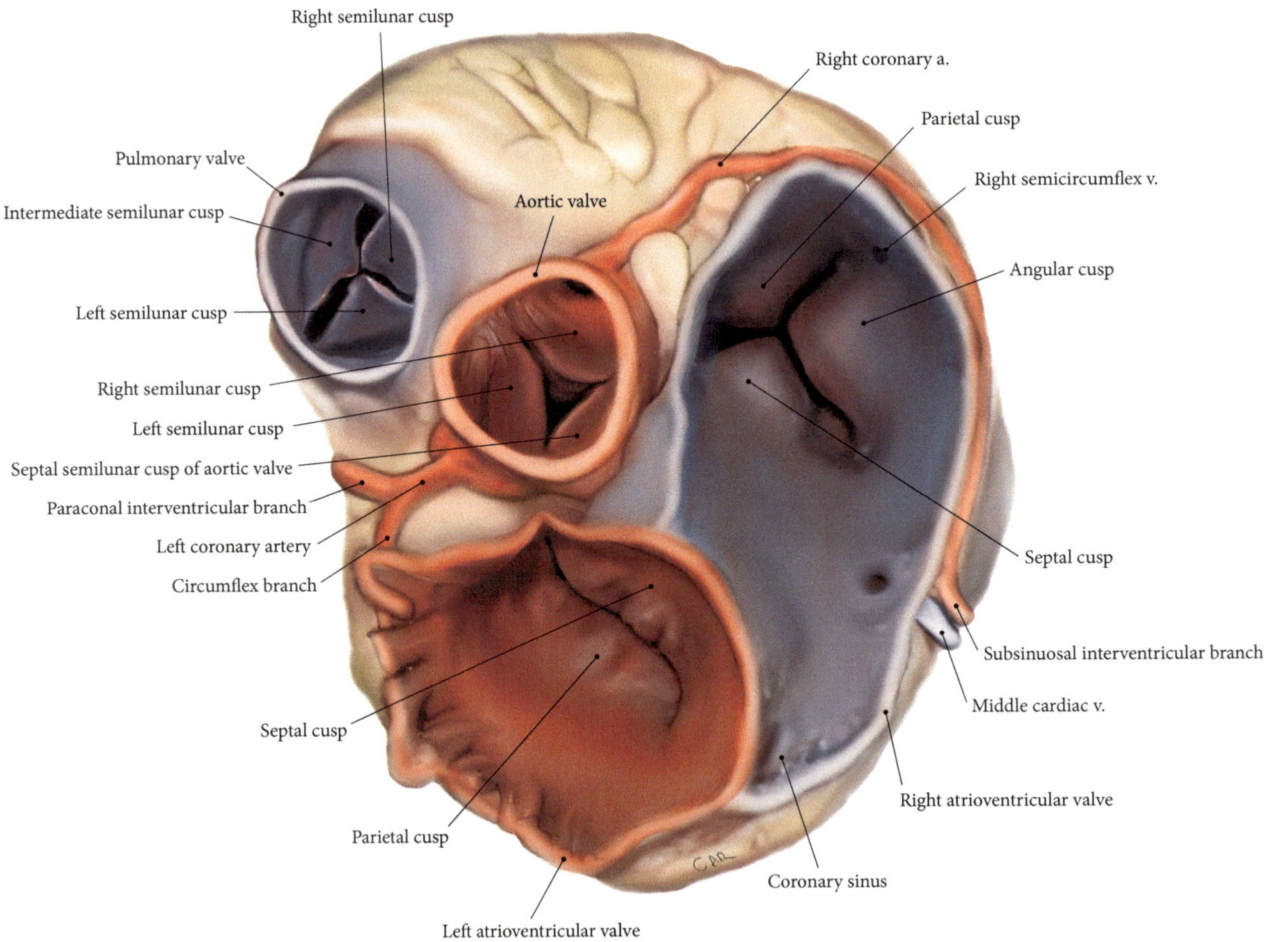

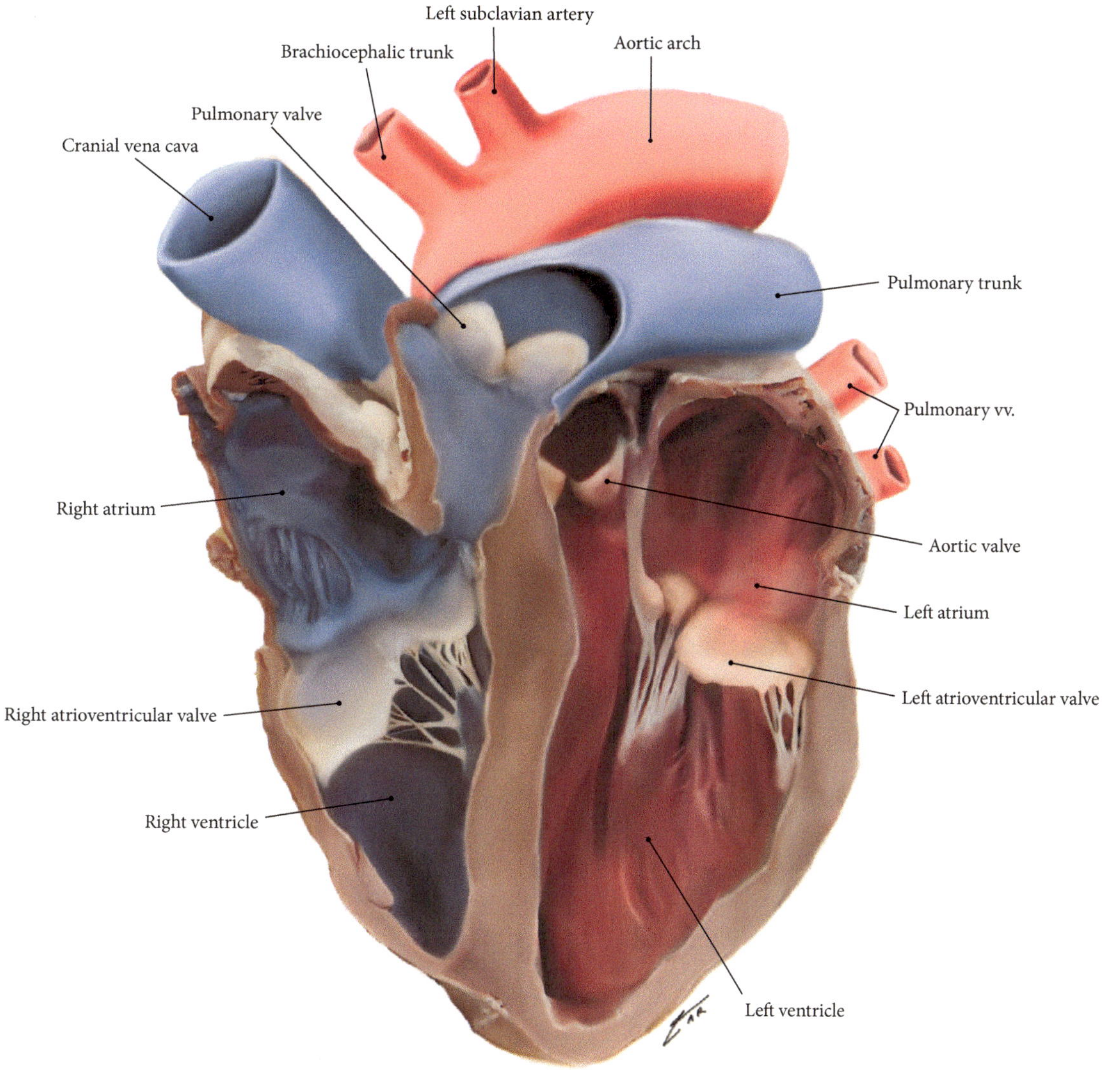
Left subclavian artery
Brachiocephalic trunk
Aortic arch
Pulmonary valve
Cranial vena cava
Pulmonary trunk
Pulmonary vv.
Right atrium
Aortic valve
Left atrium
Left atrioventricular valve
Right atrioventricular valve
Right ventricle
Left ventricle

Myocardial vasculature

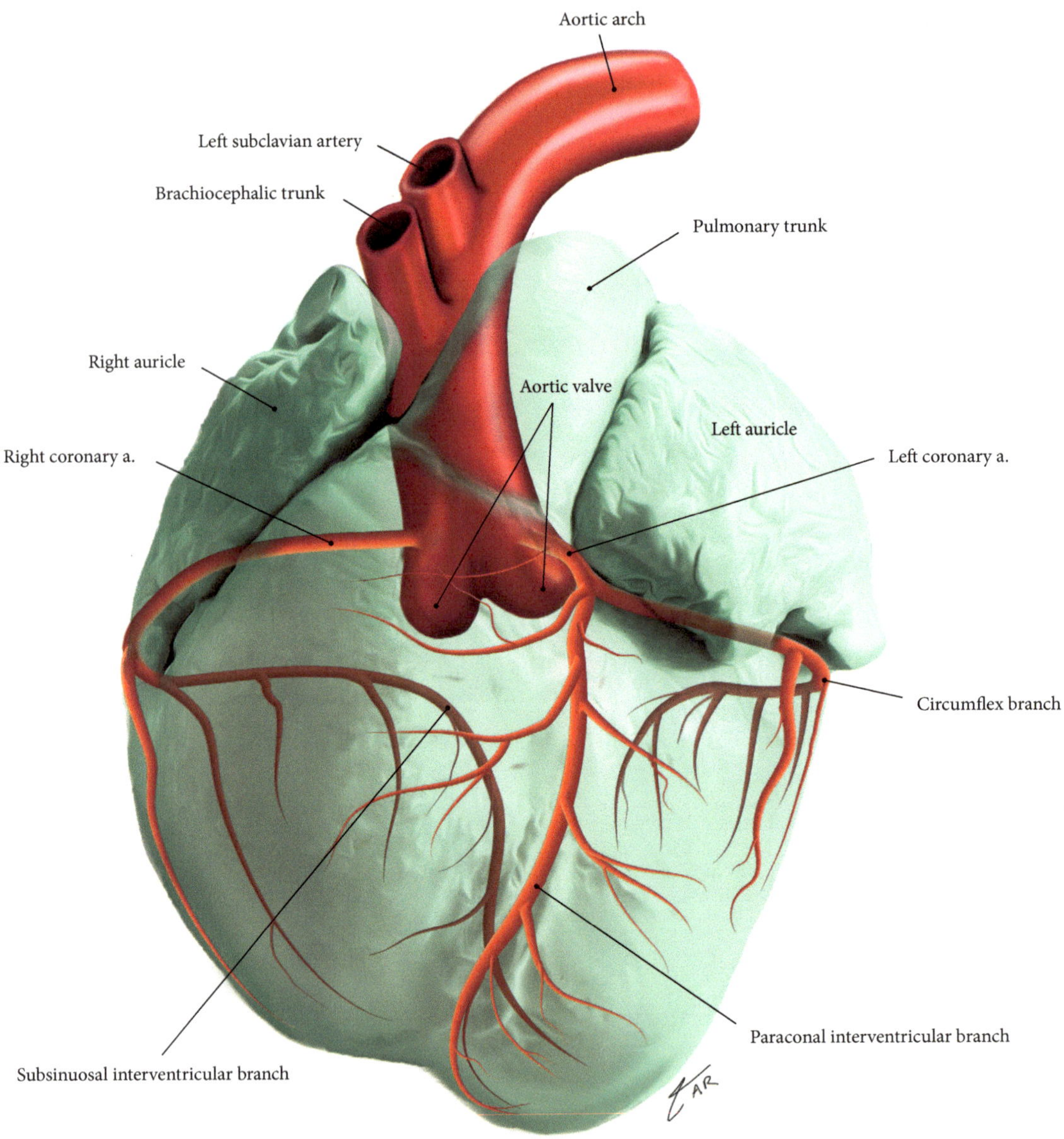

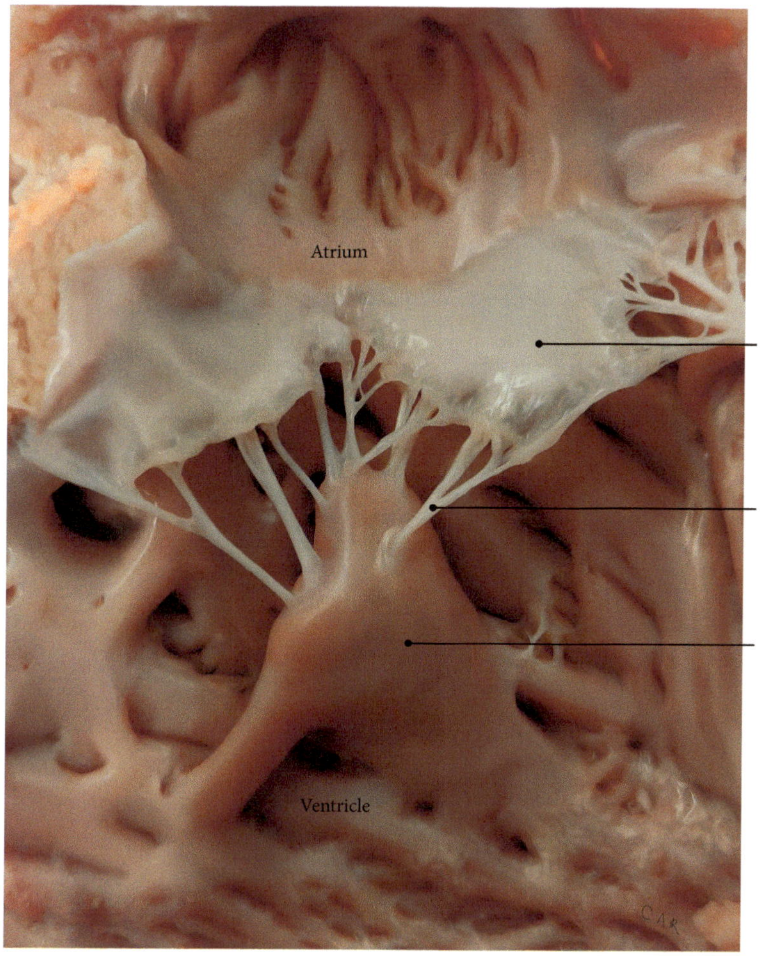
Atrium
Cusp of right atrioventricular valve
Chordae tendinae
Papillary muscles
Ventricle

Conduction systems of the heart

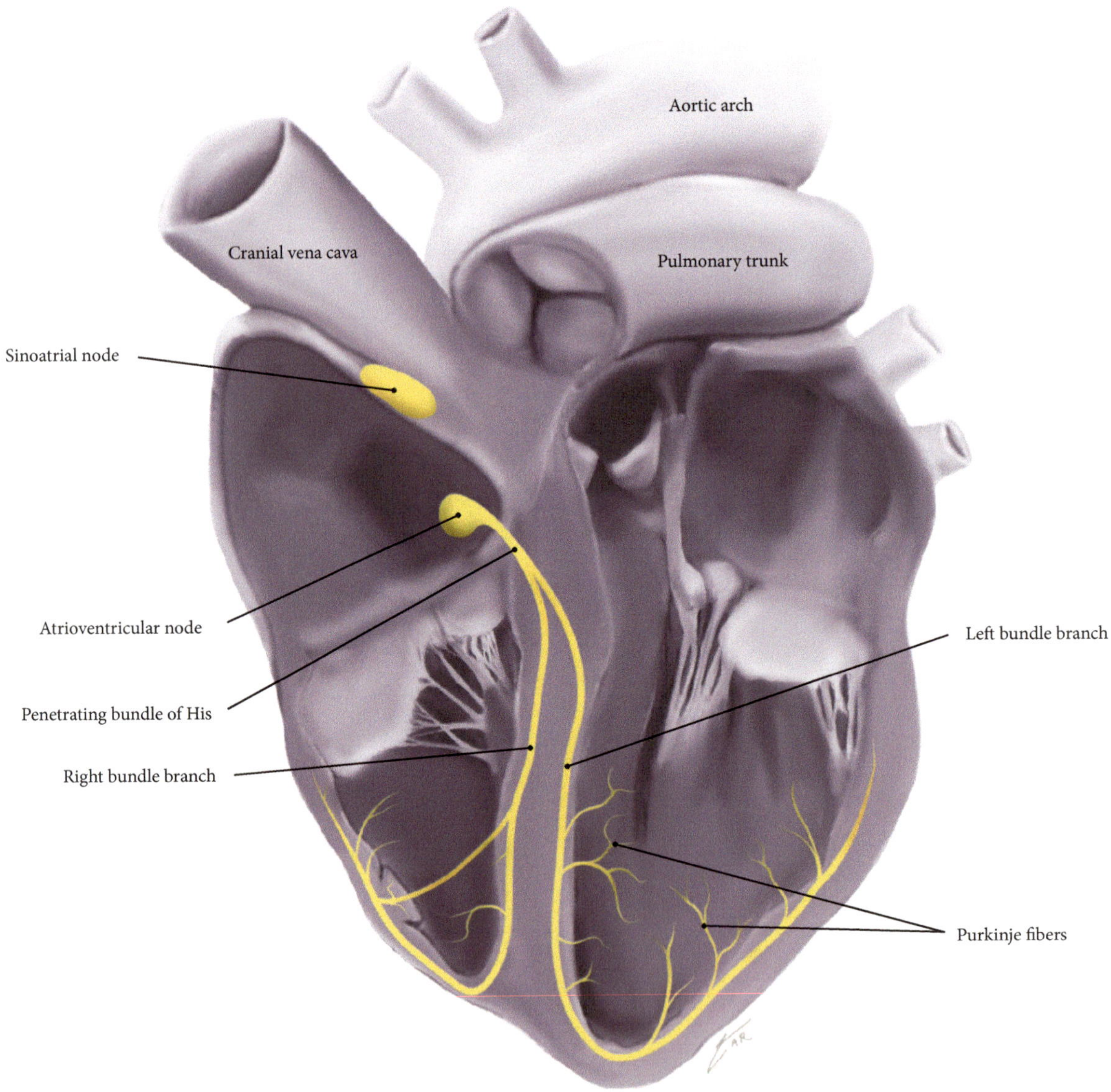

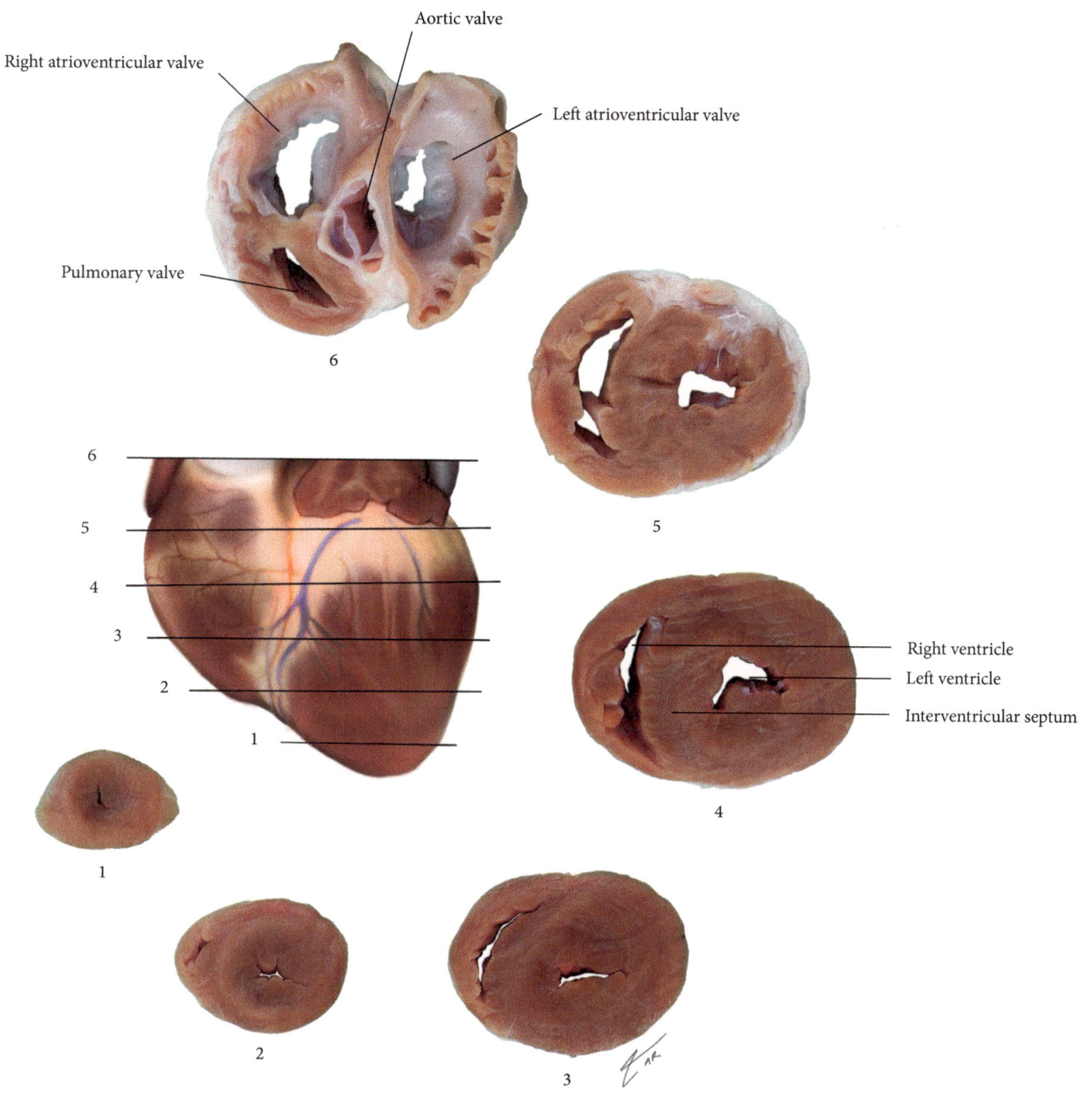
Aortic valve
Right atrioventricular valve
Left atrioventricular valve
Pulmonary valve
6
5
4
3
2
1
Right ventricle
Left ventricle
Interventricular septum

Ventral view of pig thoracic cavity – veins

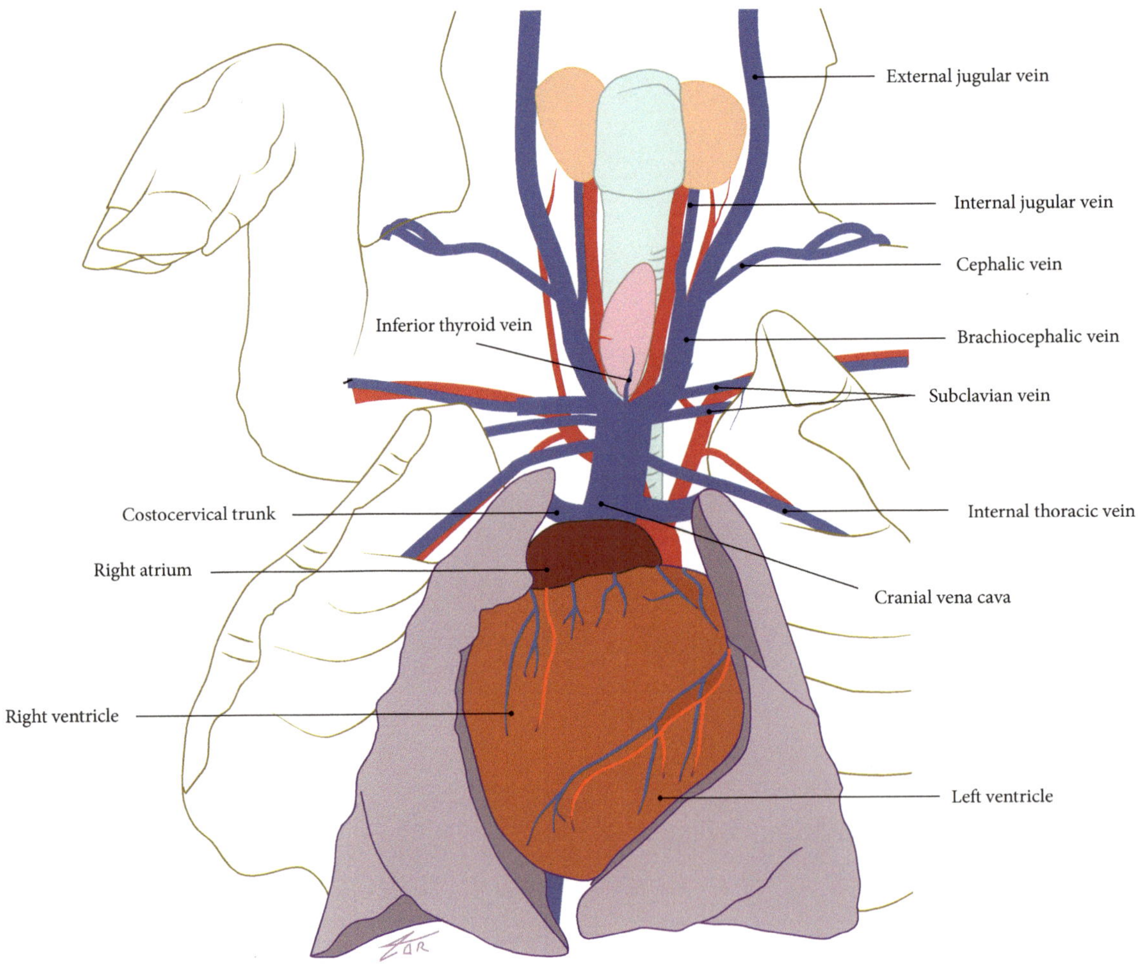

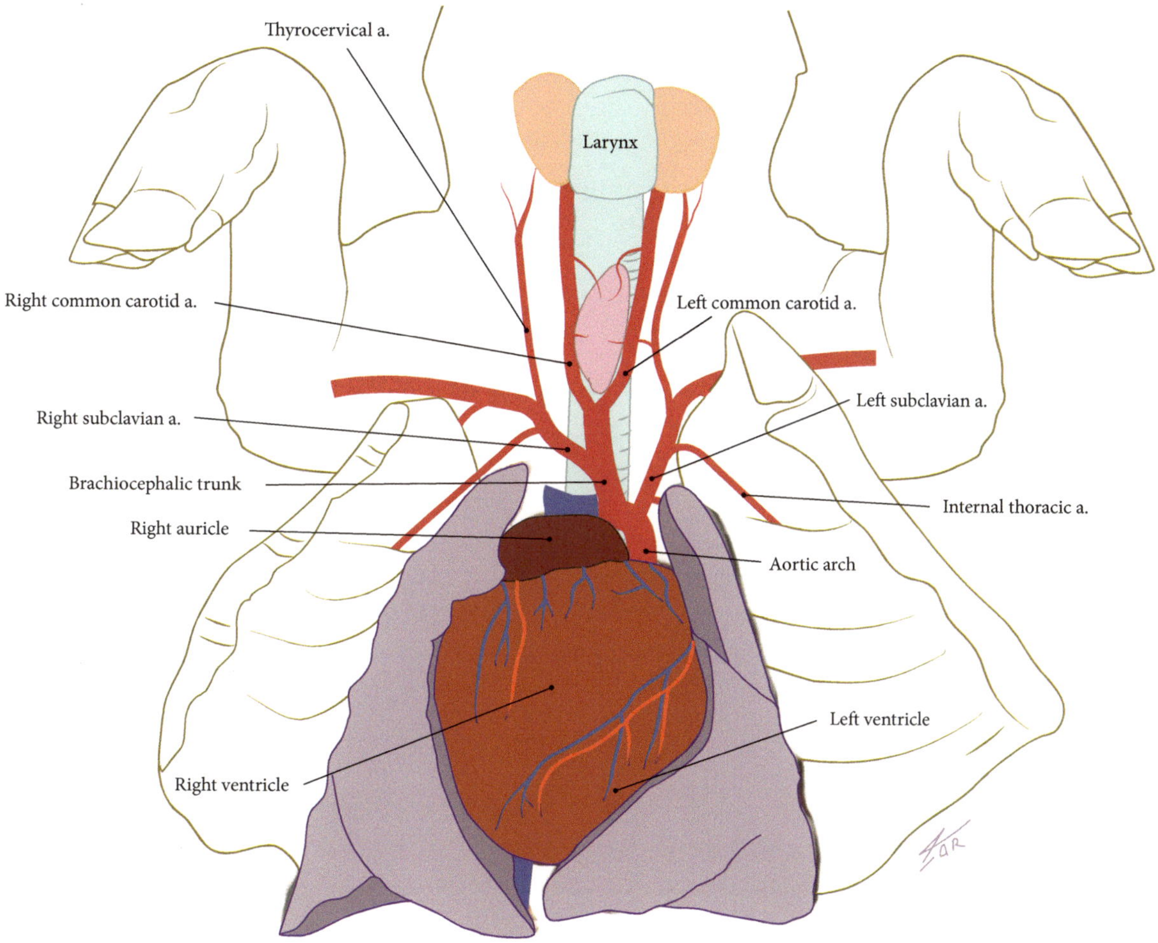
Thyrocervical a.
Larynx
Right common carotid a.
Left common carotid a.
Left subclavian a.
Right subclavian a.
Brachiocephalic trunk
Internal thoracic a.
Right auricle
Aortic arch
Left ventricle
Right ventricle

Arteries of the trunk

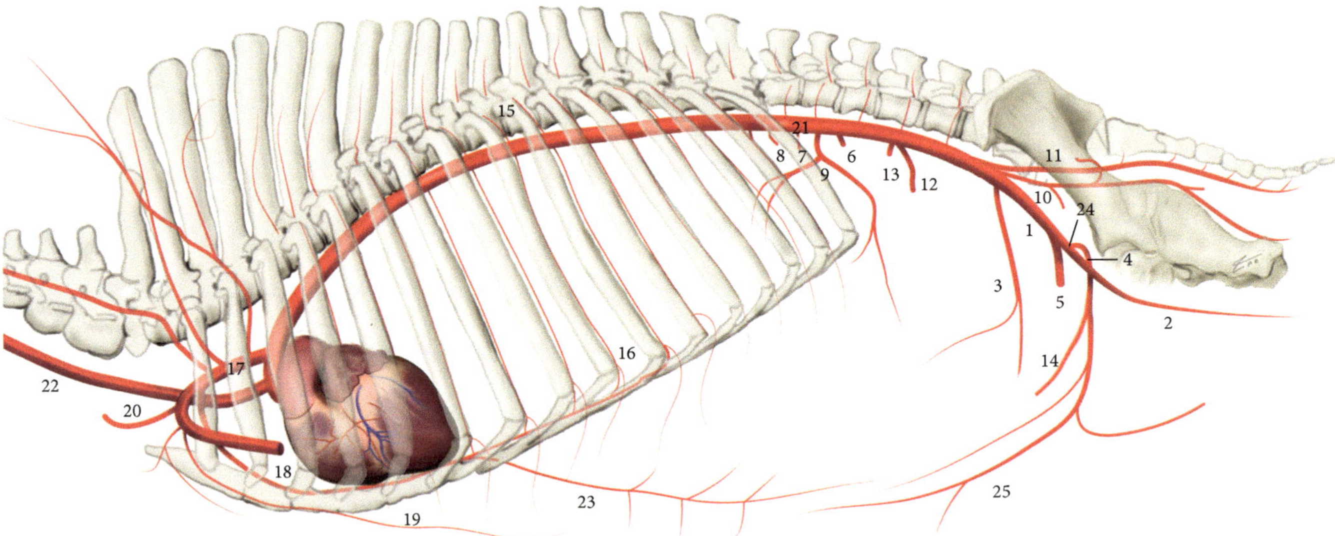

1. External iliac a.
2. Medial circumflex femoral a.
3. Deep circumflex iliac a.
4. Pudendoepigastric trunk
5. Femoral a.
6. Renal a.
7. Cranial mesenteric a.
8. Celiac a.
9. Cranial abdominal a.
10. Internal iliac a.
11. Median sacral a.
12. Caudal mesenteric a.
13. Ovarian a.
14. Caudal epigastric a.
15. Dorsal intercostal aa.
16. Ventral intercostal aa.
17. Subclavian a.
18. Internal thoracic a.
19. External thoracic a.
20. Superficial cervical a.
21. Abdominal aorta
22. Common carotid a.
23. Cranial epigastric a.
24. Deep femoral a.
25. External pudendal a.

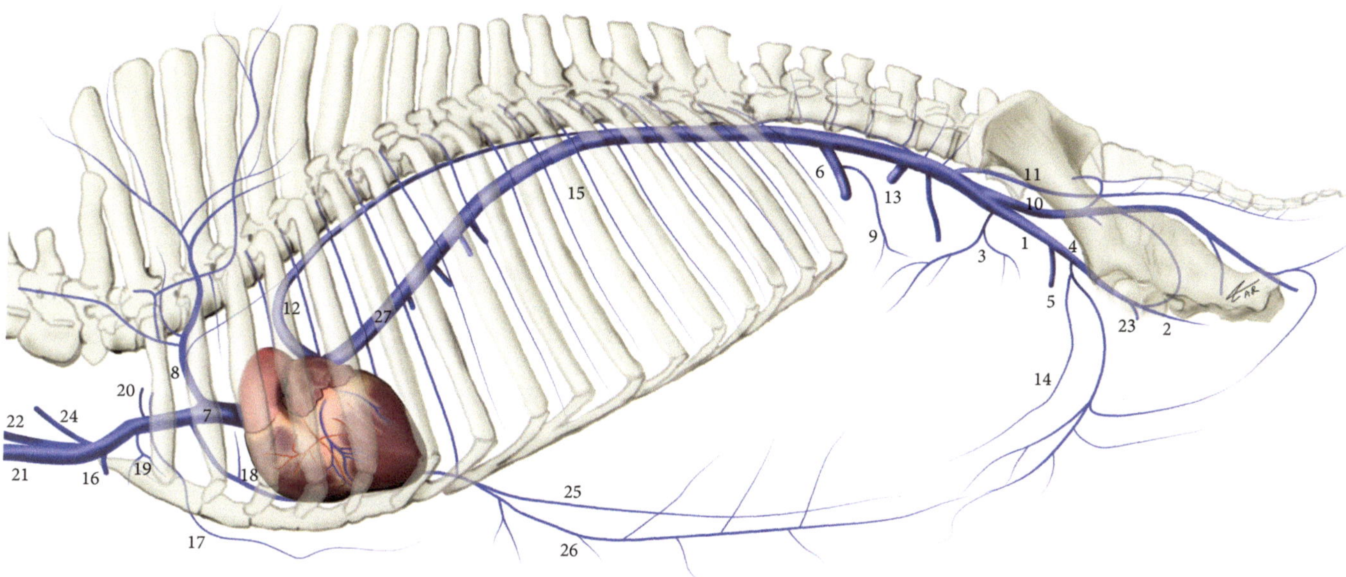

1. External iliac v.
2. Medial circumflex femoral v.
3. Deep circumflex iliac v.
4. Pudendoepigastric v.
5. Femoral v.
6. Renal v.
7. Cranial vena cava
8. Costocervical v.
9. Cranial abdominal v.
10. Internal iliac v.
11. Median sacral v.
12. Azygos v.
13. Ovarian/Testicular v.
14. Caudal epigastric v.
15. Dorsal intercostal v.
16. Cephalic v.
17. Lateral thoracic v.
18. Internal thoracic v.
19. External thoracic v.
20. Proximal axillary v.
21. External jugular v.
22. Internal jugular v.
23. Deep femoral v.
24. Superficial cervical v.
25. Cranial epigastric v.
26. Superficial cranial epigastric v.
27. Caudal vena cava

Thyroid gland and associated blood vessels

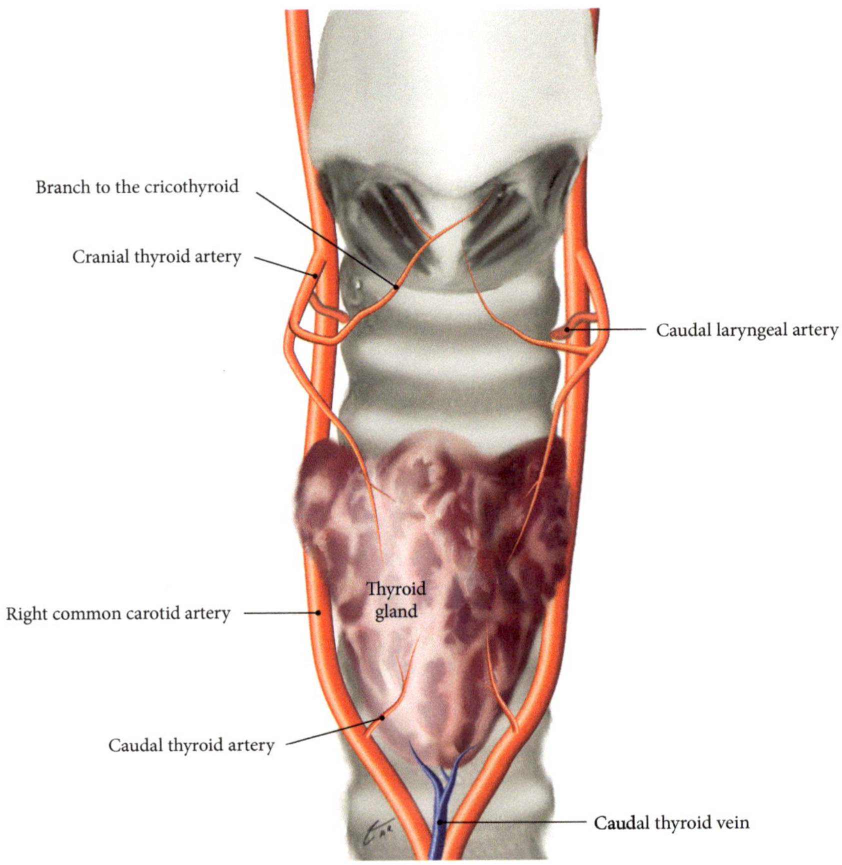

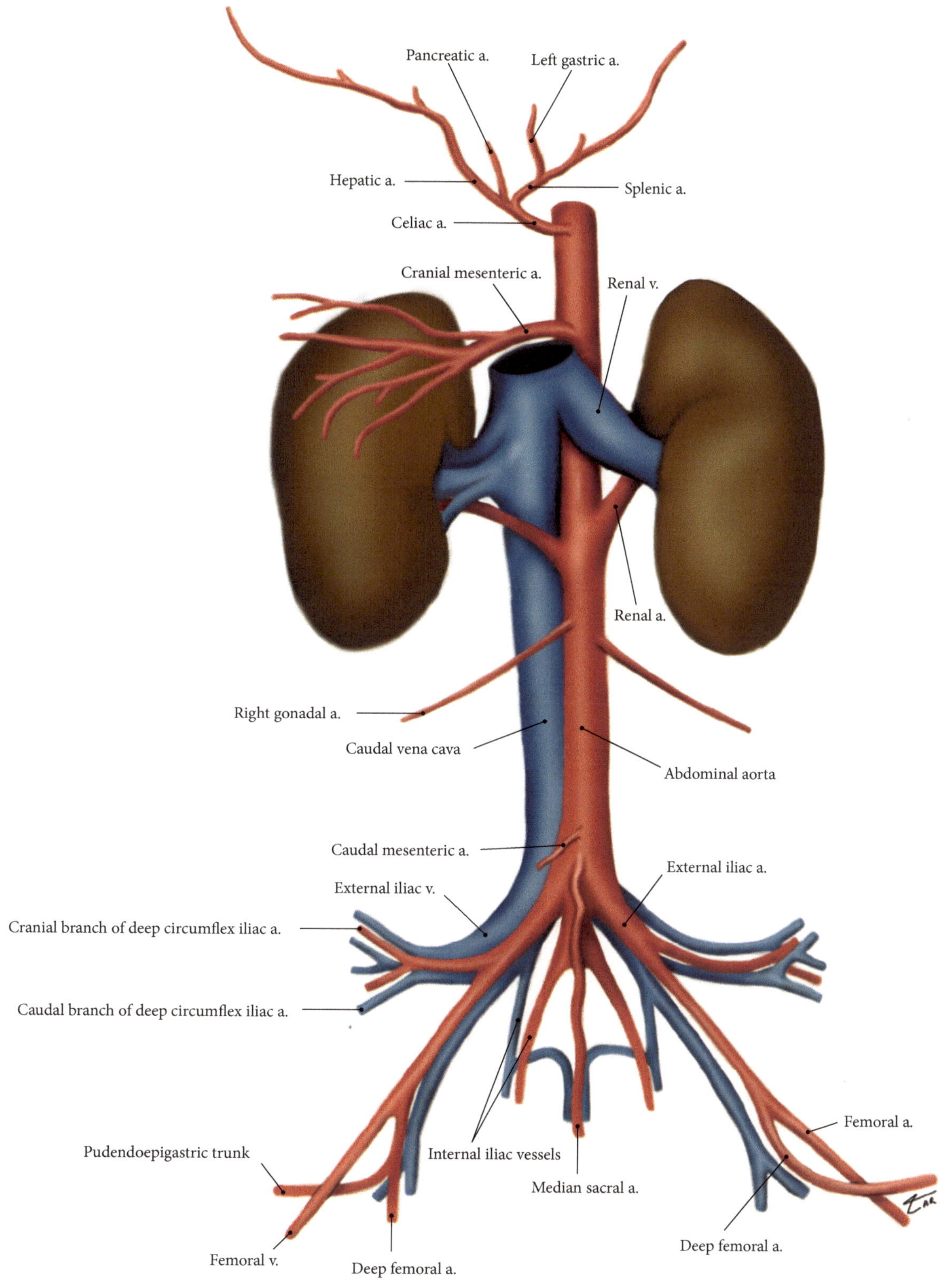
Pancreatic a.
Left gastric a.
Hepatic a.
Splenic a.
Celiac a.
Cranial mesenteric a.
Renal v.
Renal a.
Right gonadal a.
Caudal vena cava
Abdominal aorta
Caudal mesenteric a.
External iliac a.
External iliac v.
Cranial branch of deep circumflex iliac a.
Caudal branch of deep circumflex iliac a.
Femoral a.
Pudendoepigastric trunk
Internal iliac vessels
Median sacral a.
Deep femoral a.
Femoral v.
Deep femoral a.

Arterial blood supply to thoracic limb of the pig – medial view

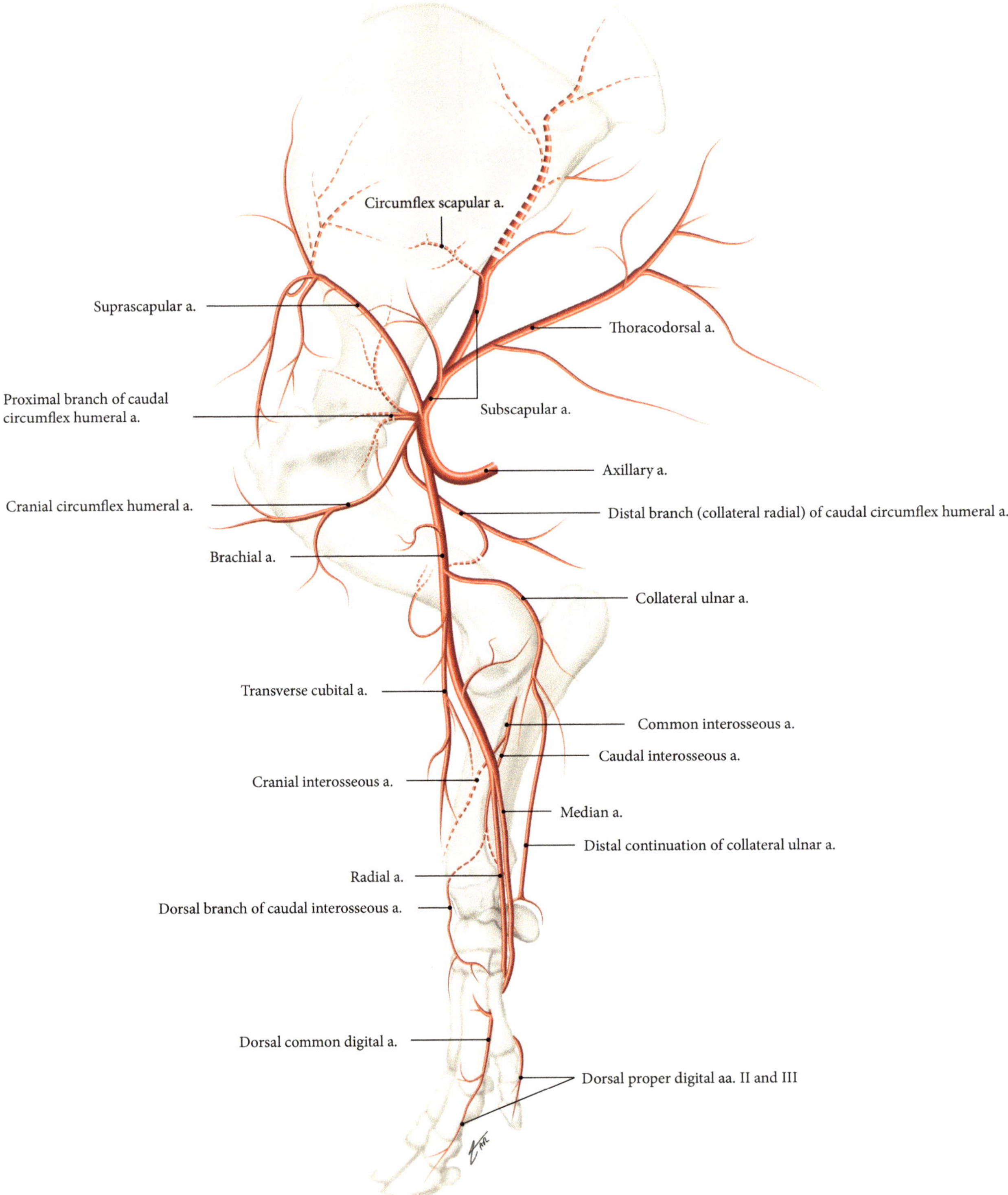

Arteries of distal part of left thoracic limb – dorsal view

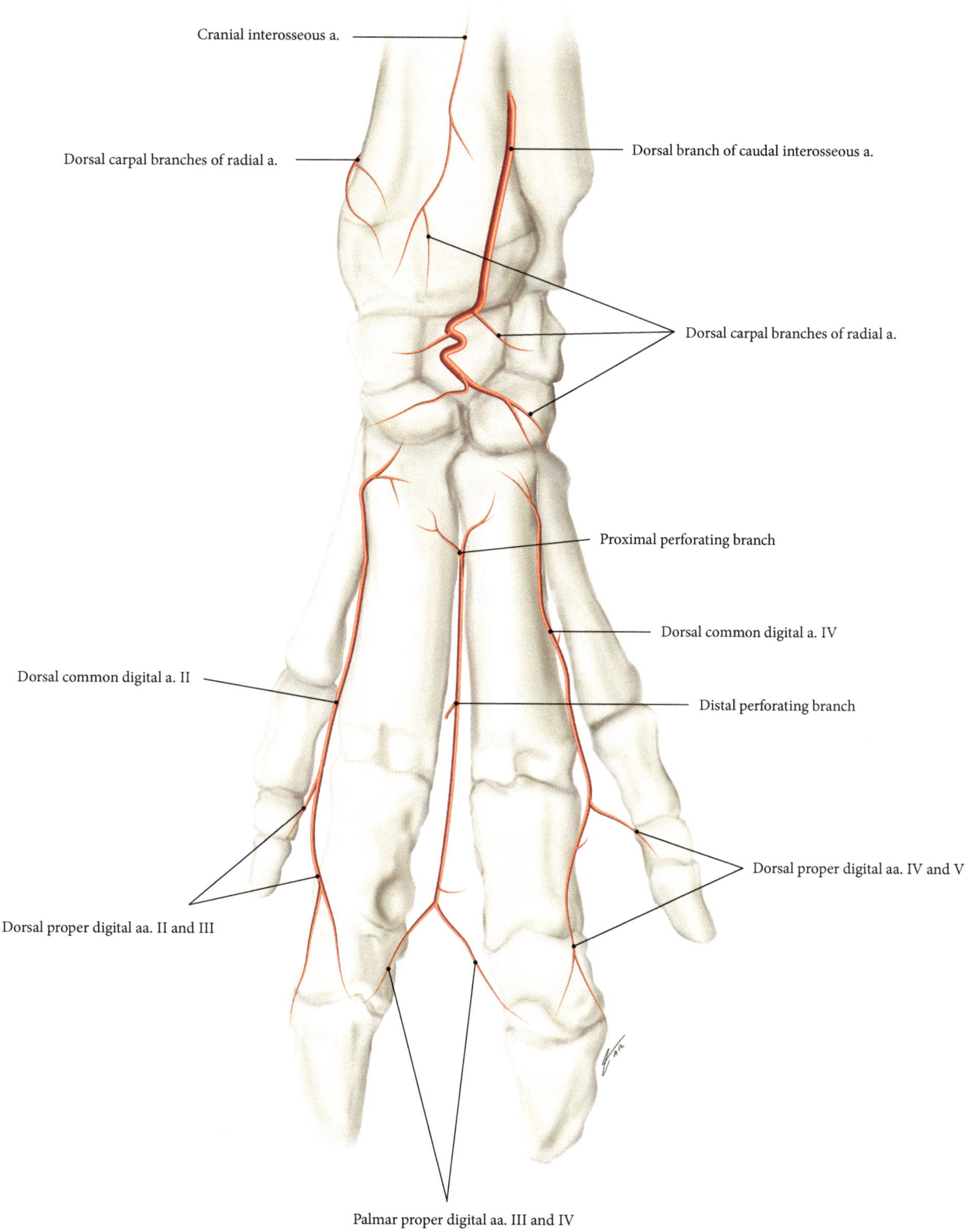

Arteries of distal part of left thoracic limb – palmar view

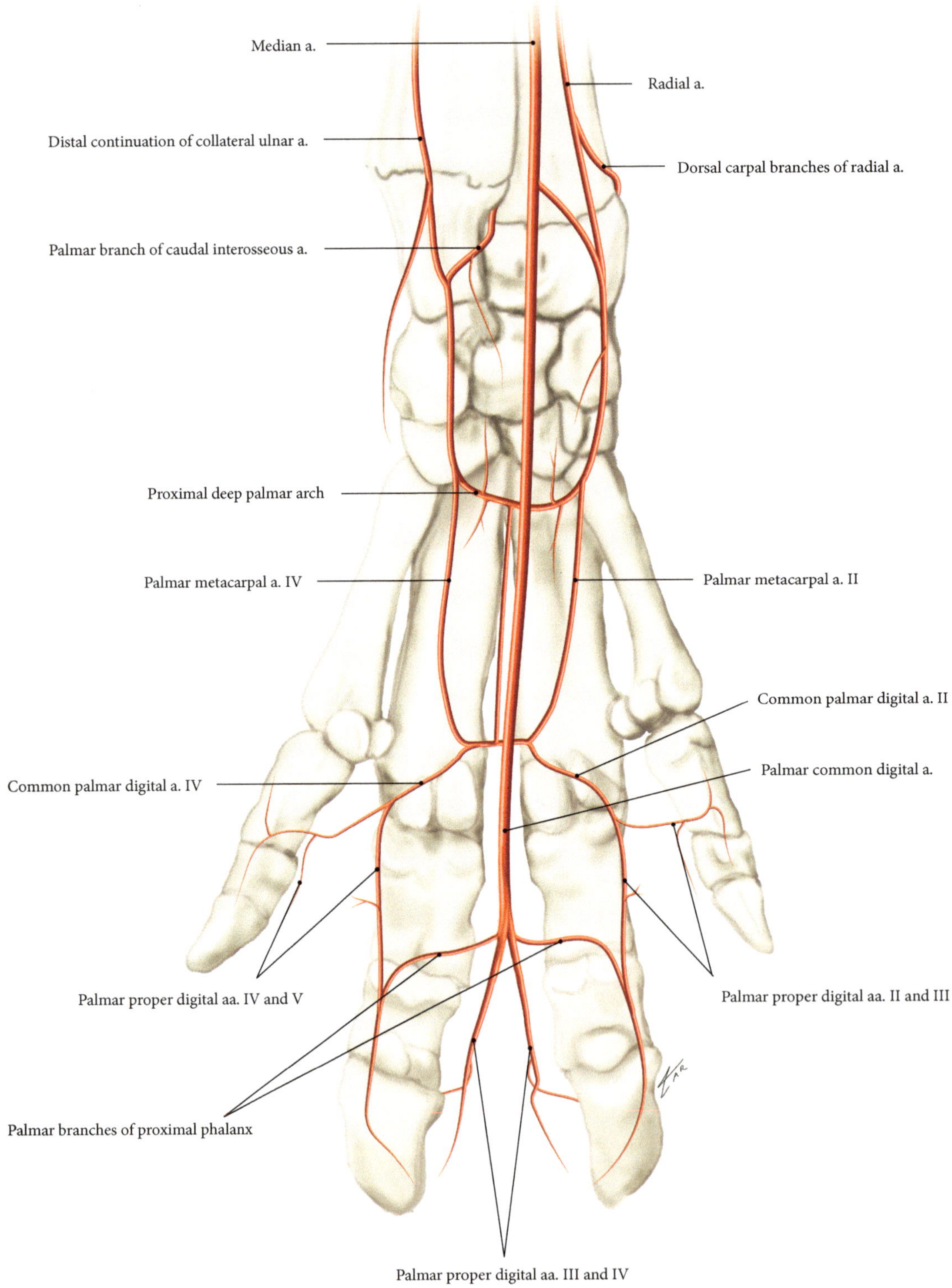

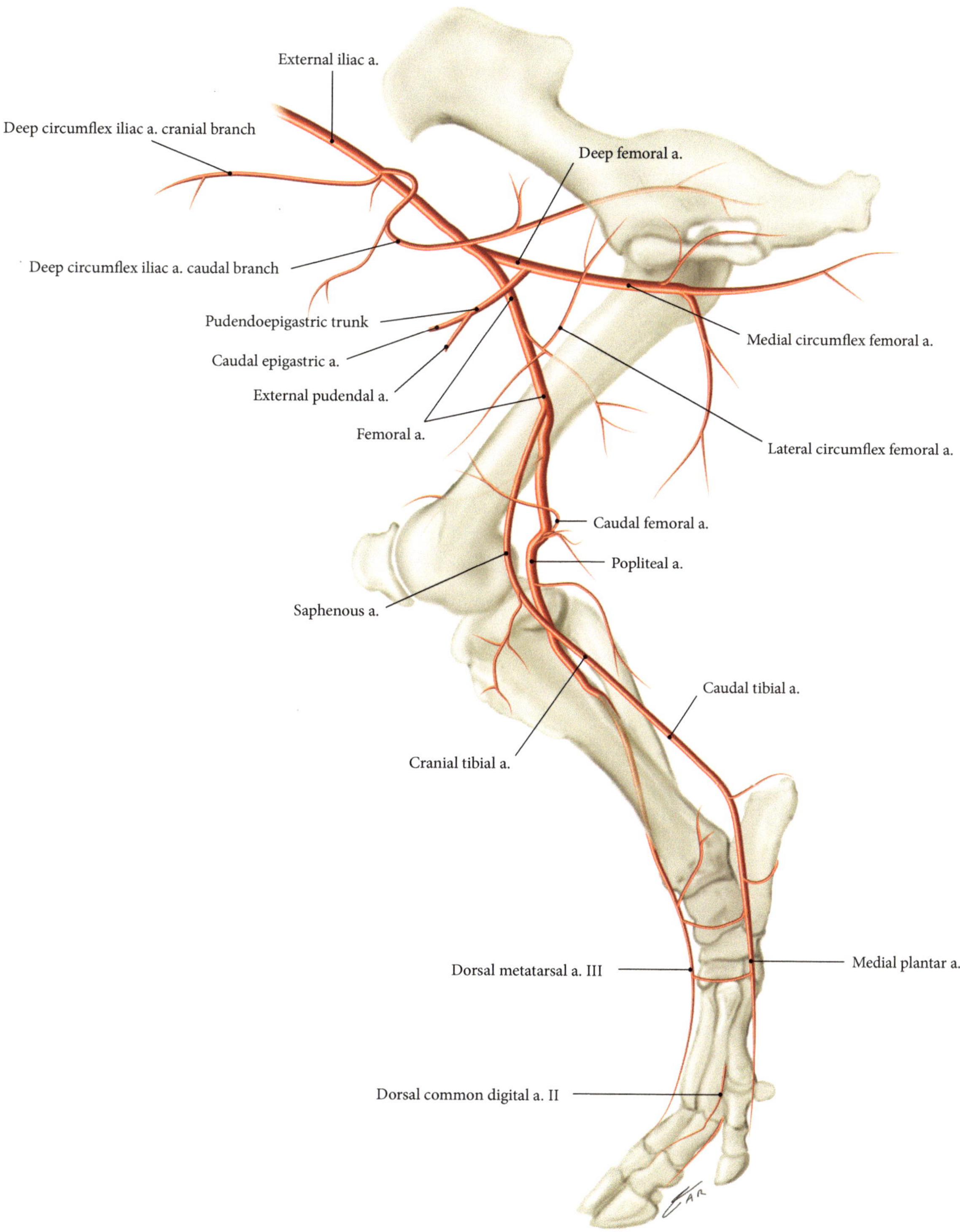
External iliac a.
Deep circumflex iliac a. cranial branch
Deep femoral a.
Deep circumflex iliac a. caudal branch
Pudendoepigastric trunk
Caudal epigastric a.
External pudendal a.
Femoral a.
Medial circumflex femoral a.
Lateral circumflex femoral a.
Caudal femoral a.
Popliteal a.
Saphenous a.
Caudal tibial a.
Cranial tibial a.
Dorsal metatarsal a. III
Medial plantar a.
Dorsal common digital a. II

Arteries of distal part of right pelvic limb – dorsal view

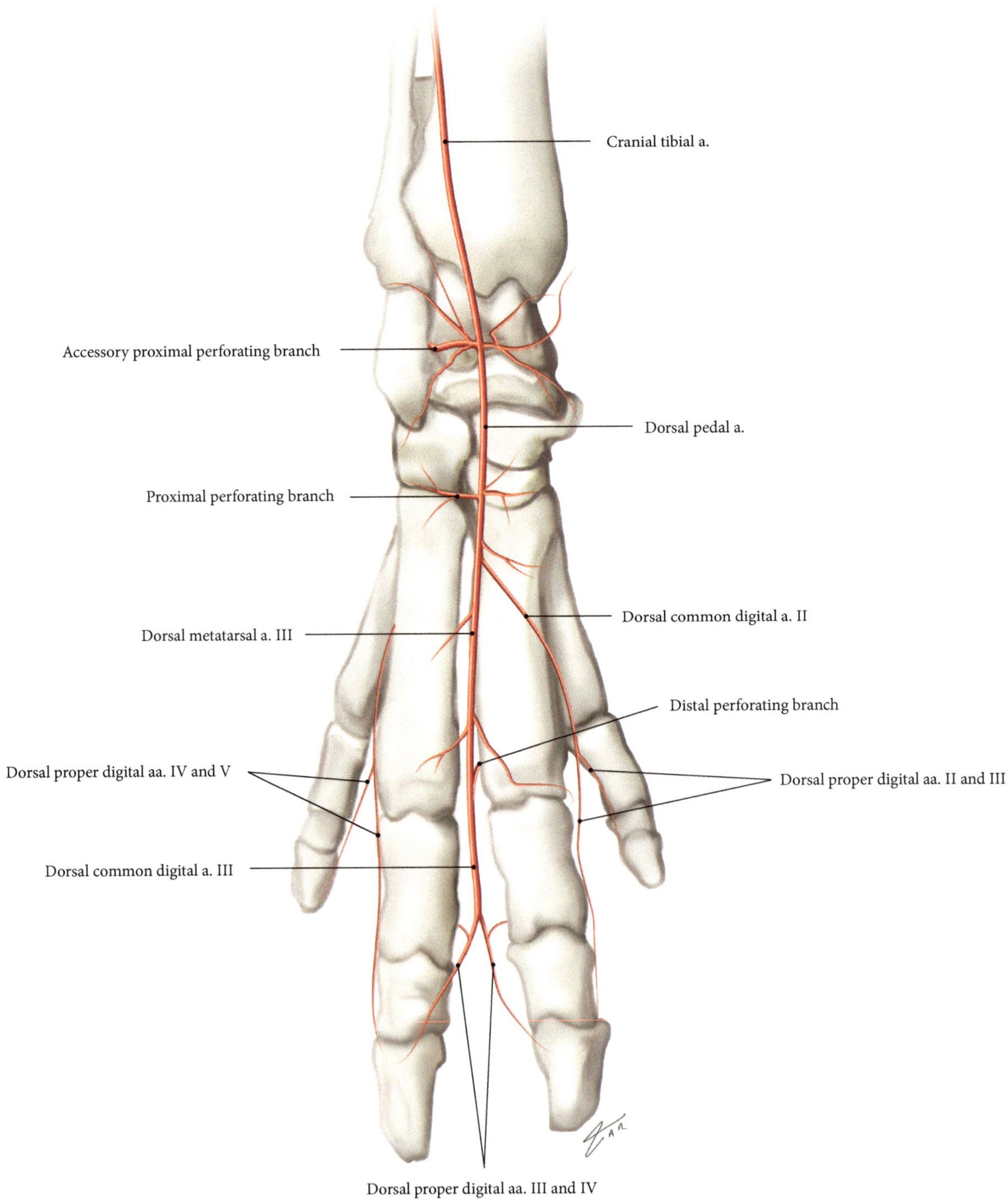

Arteries of distal part of right pelvic limb – plantar view

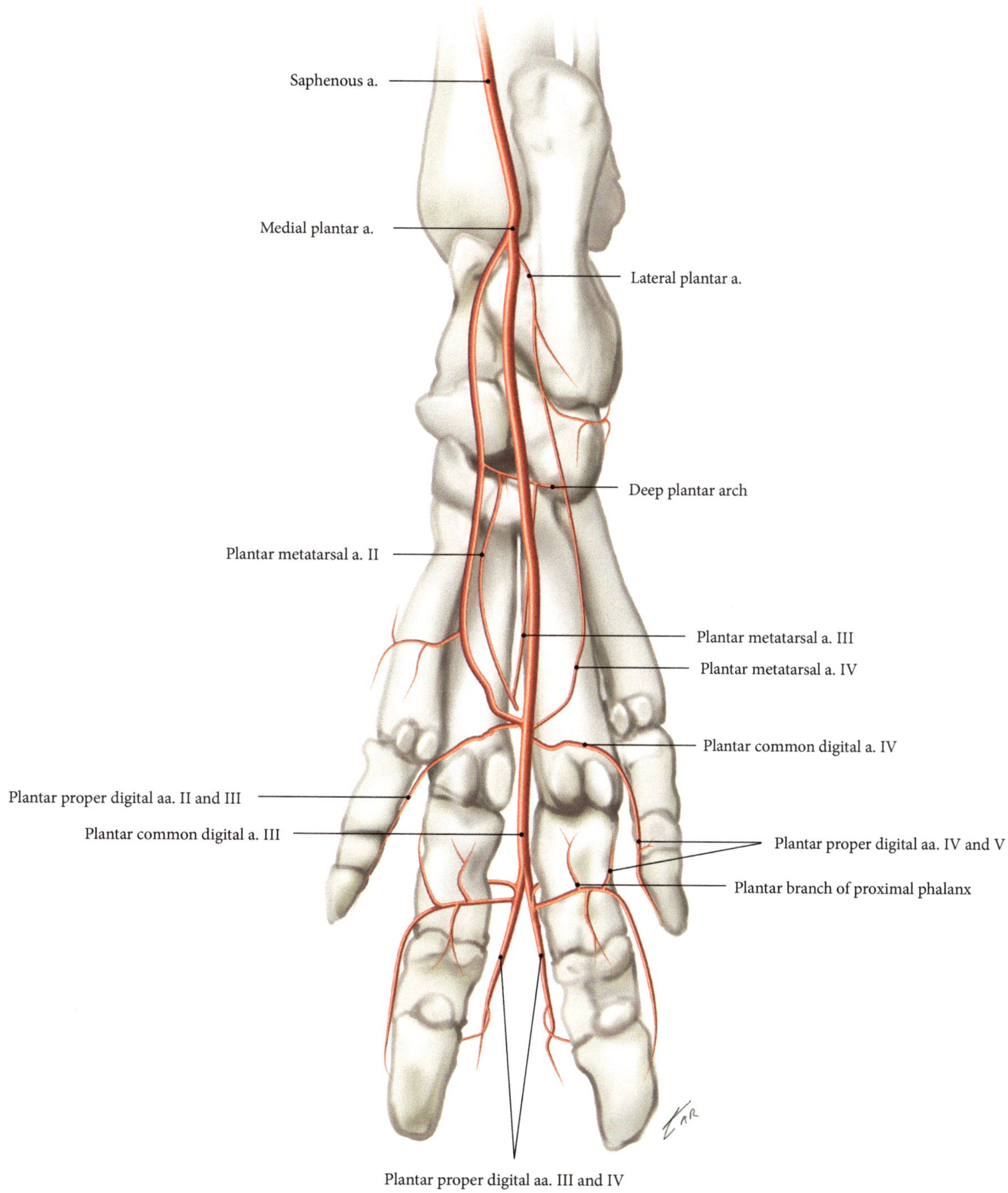

The cephalic vein

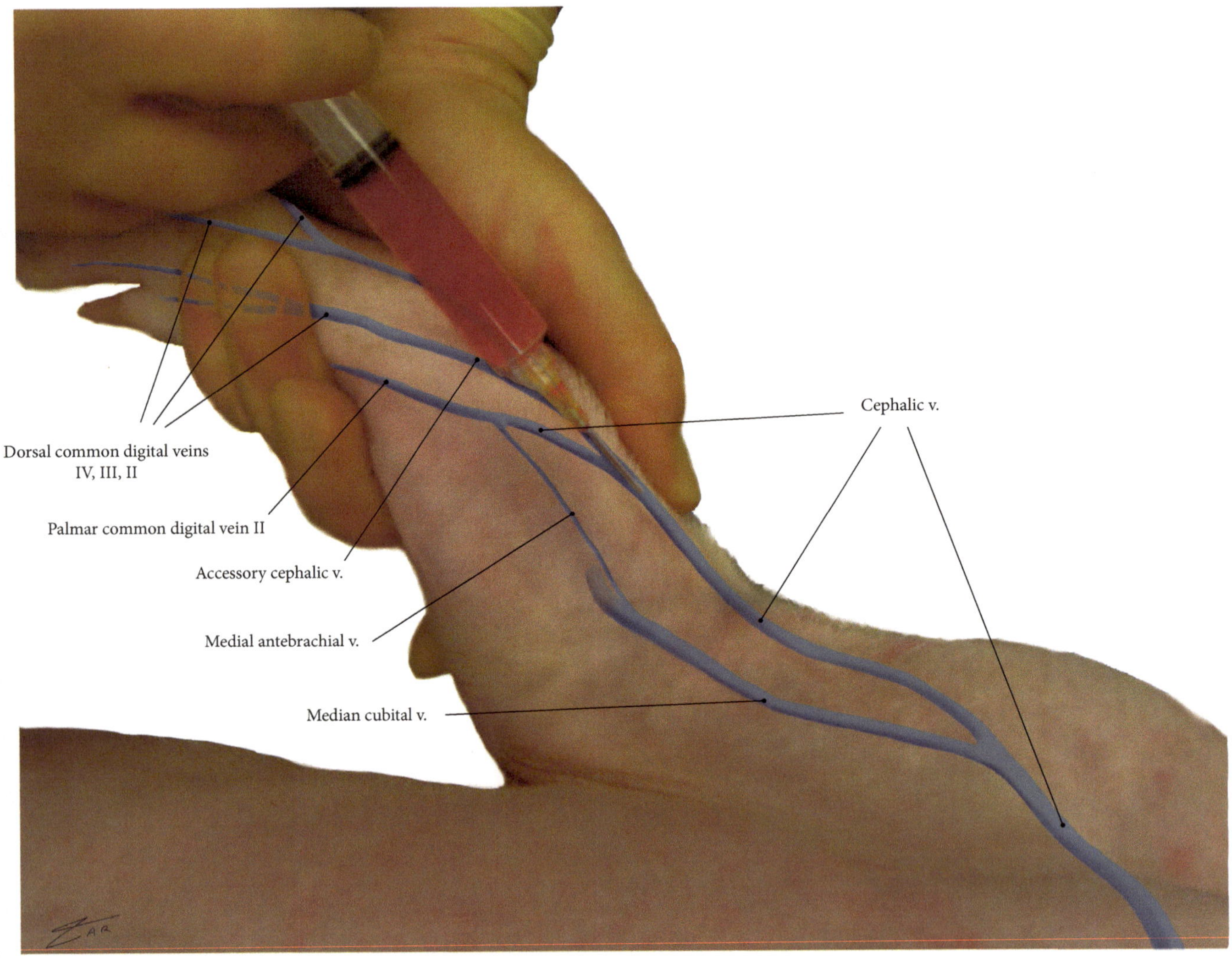

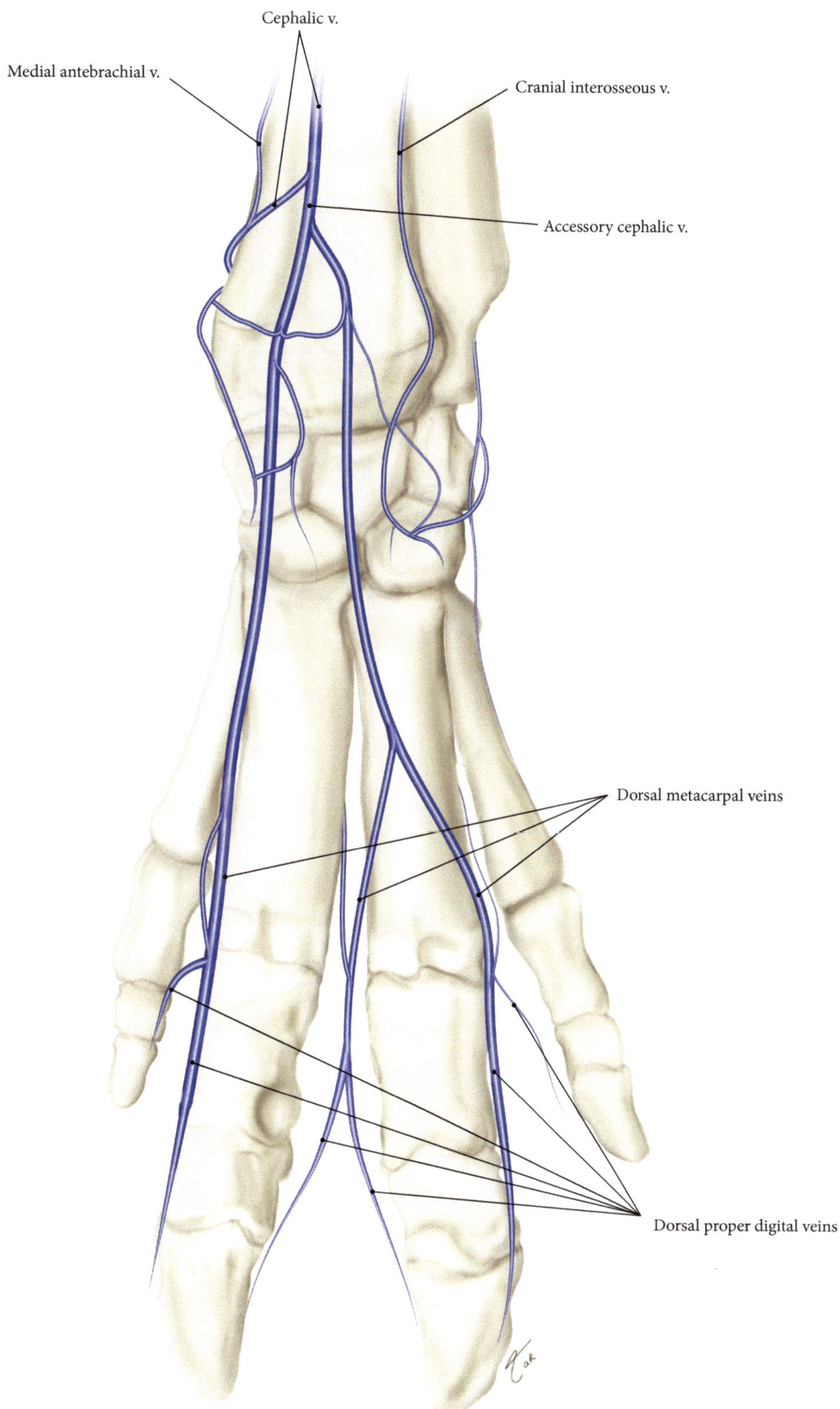
Cephalic v.
Medial antebrachial v.
Cranial interosseous v.
Accessory cephalic v.
Dorsal metacarpal veins
Dorsal proper digital veins

Veins of the thoracic autopodium – palmar view

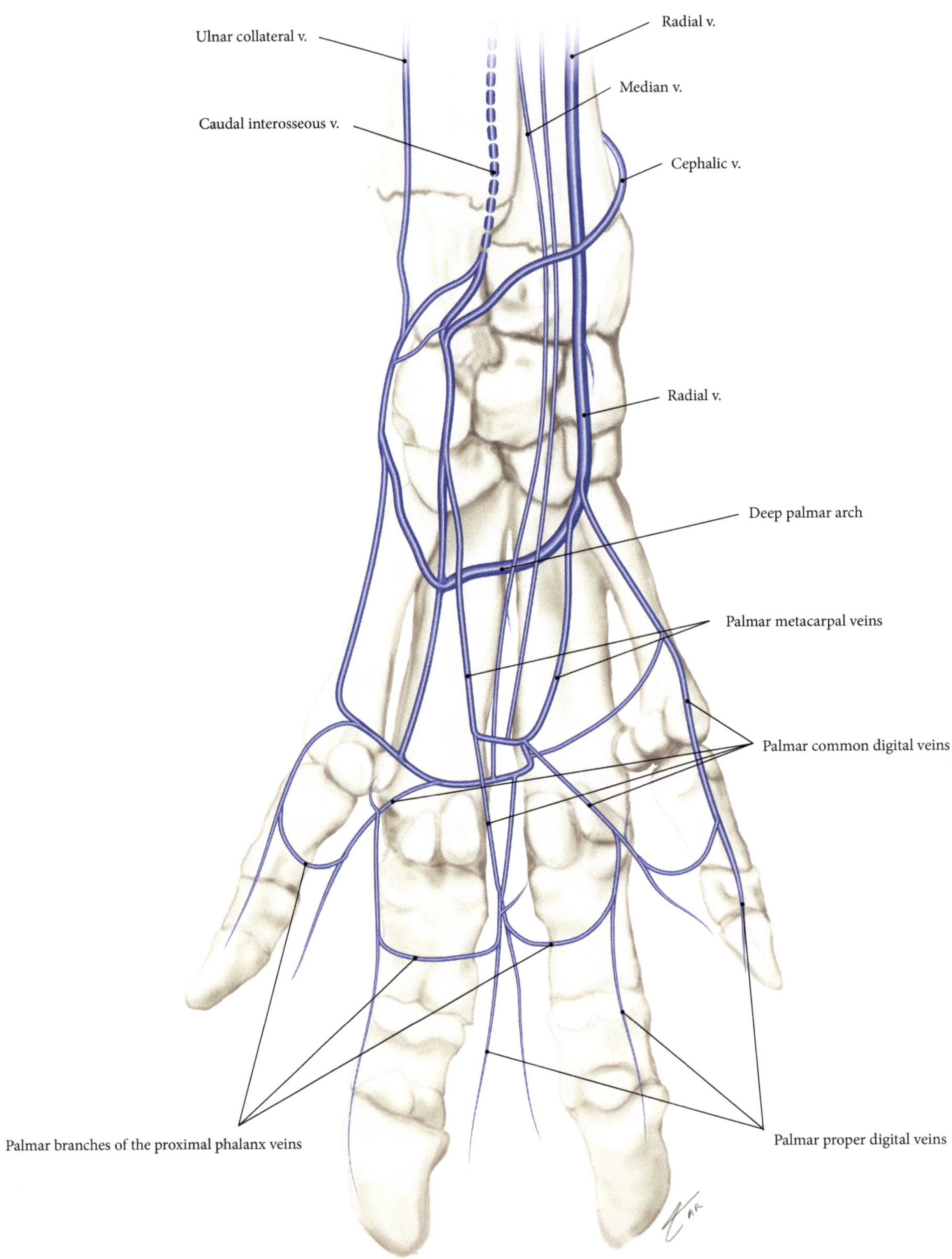

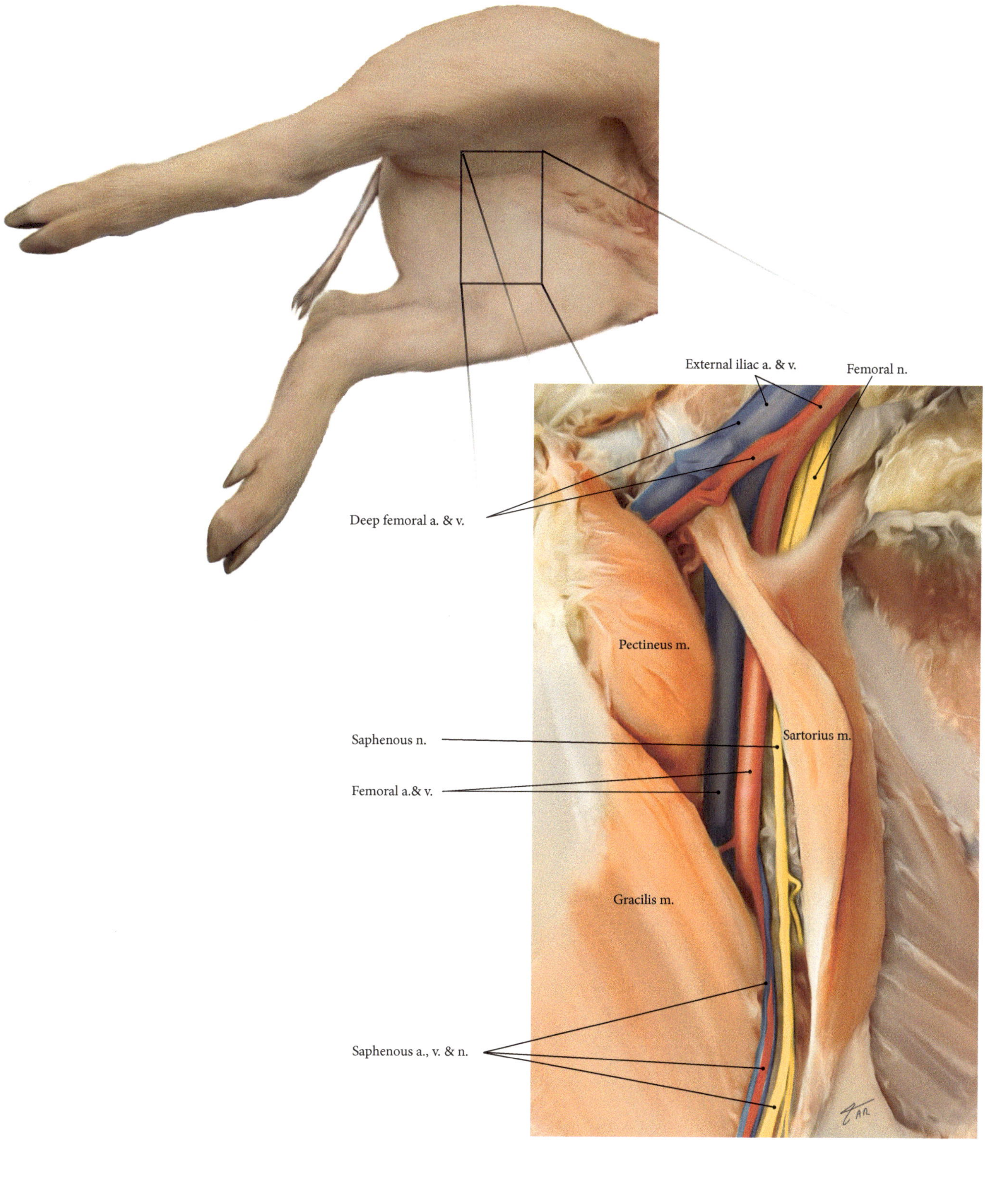
External iliac a. & v.
Femoral n.
Deep femoral a. & v.
Pectineus m.
Saphenous n.
Sartorius m.
Femoral a.& v.
Gracilis m.
Saphenous a., v. & n.

Nerves of the left hindlimb

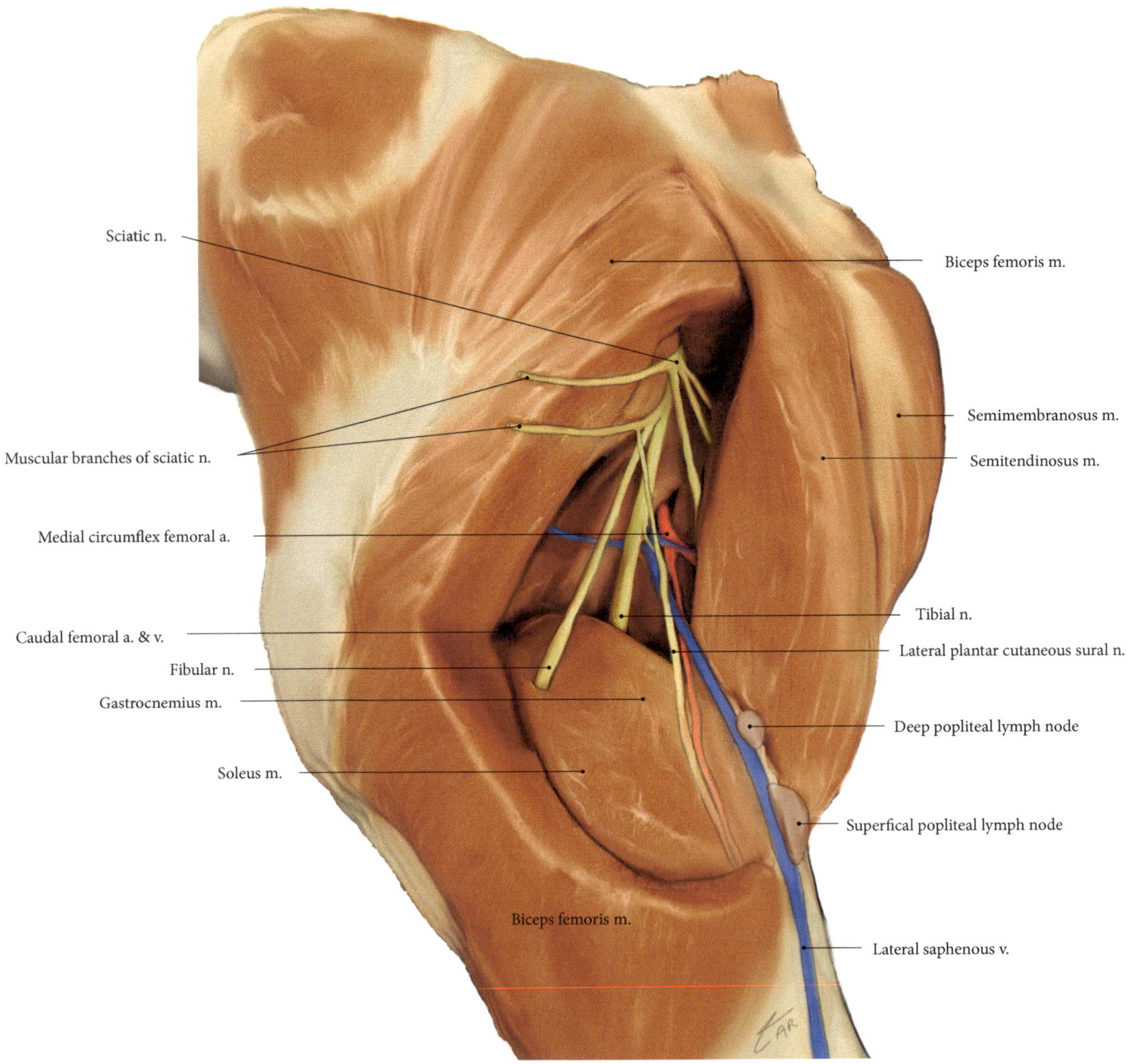

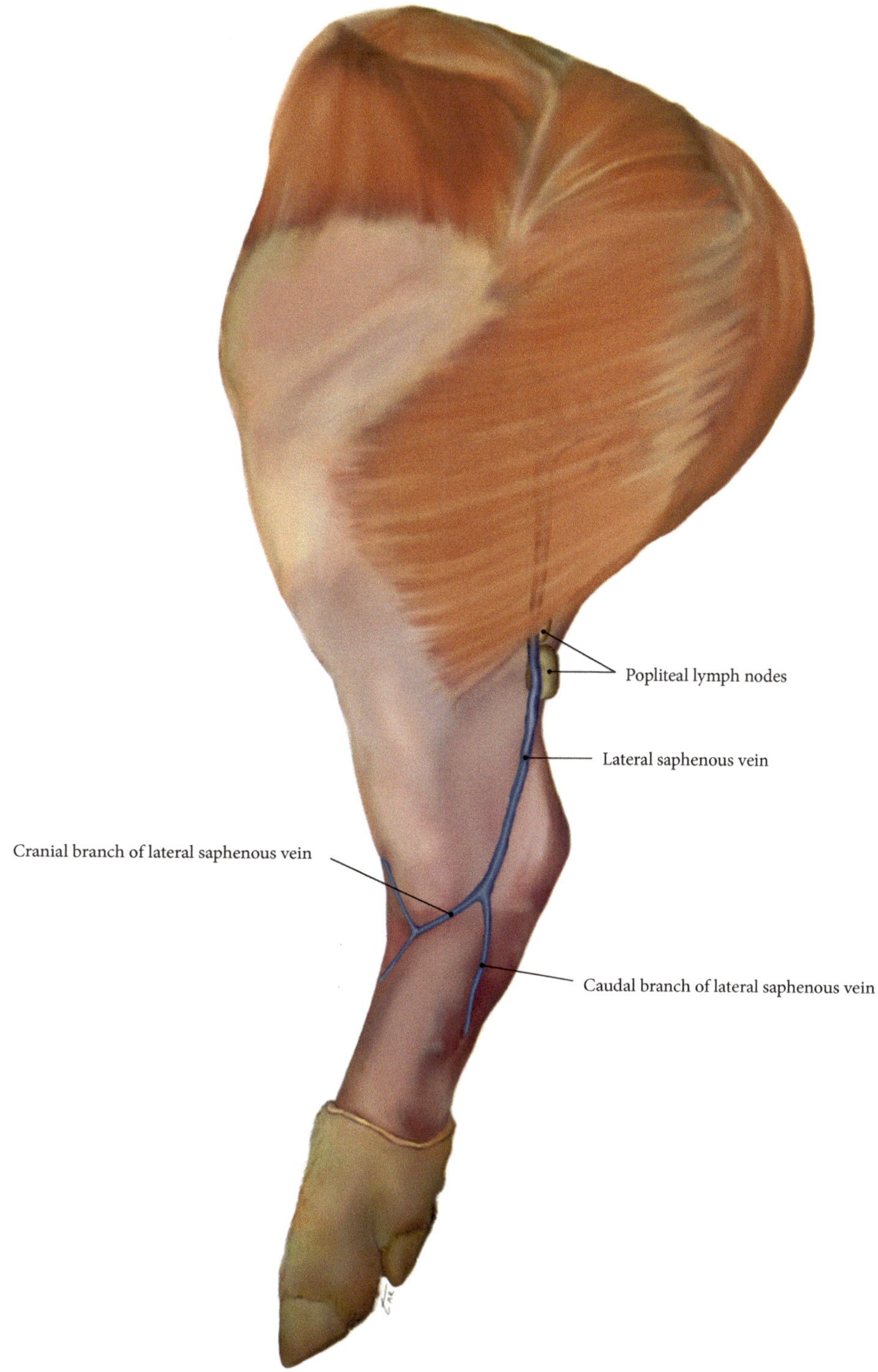
Popliteal lymph nodes
Lateral saphenous vein
Cranial branch of lateral saphenous vein
Caudal branch of lateral saphenous vein

Veins of the pelvic autopodium – dorsal view

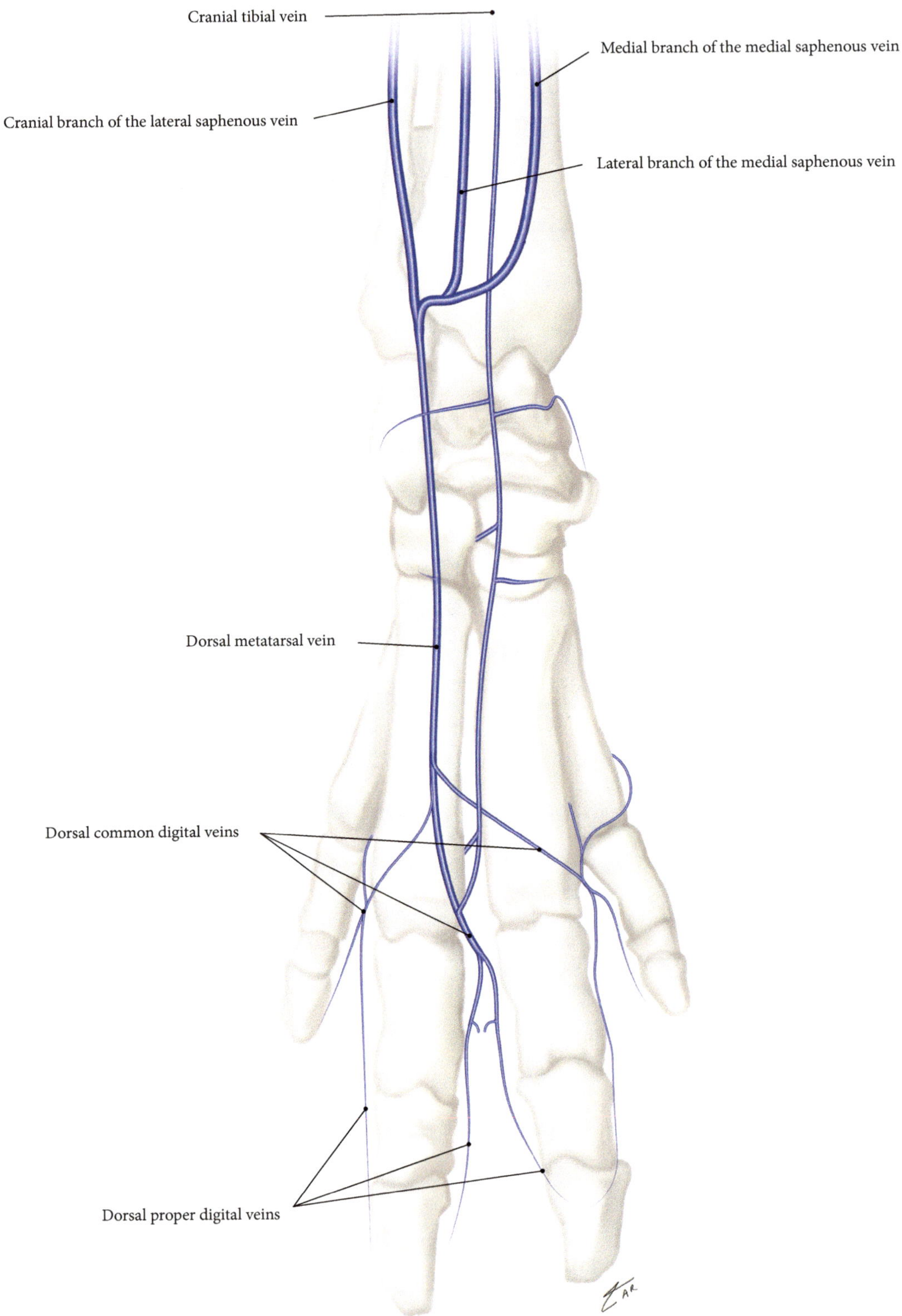

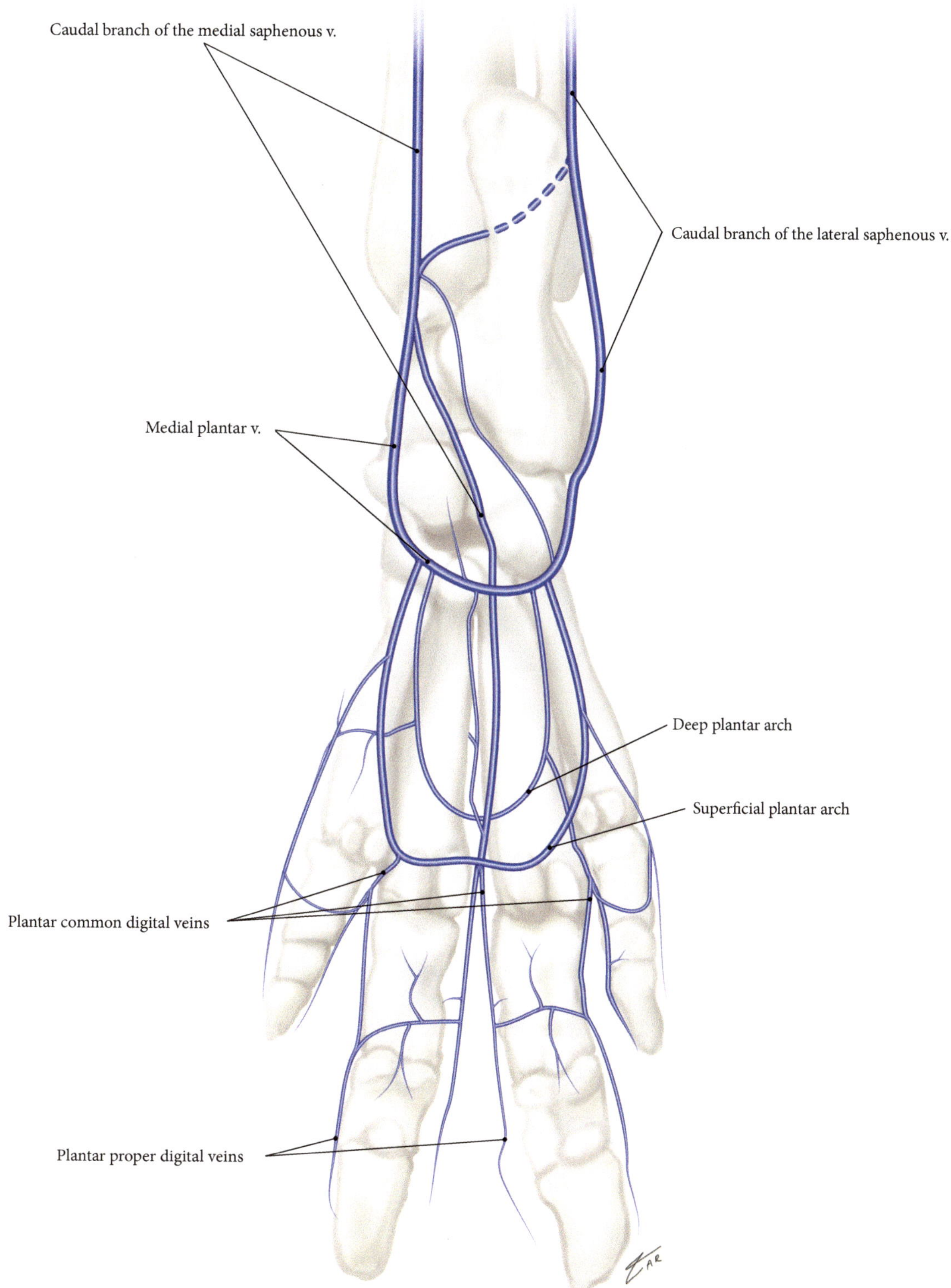
Caudal branch of the medial saphenous v.
Caudal branch of the lateral saphenous v.
Medial plantar v.
Deep plantar arch
Superficial plantar arch
Plantar common digital veins
Plantar proper digital veins

Porcine brain – ventral aspect

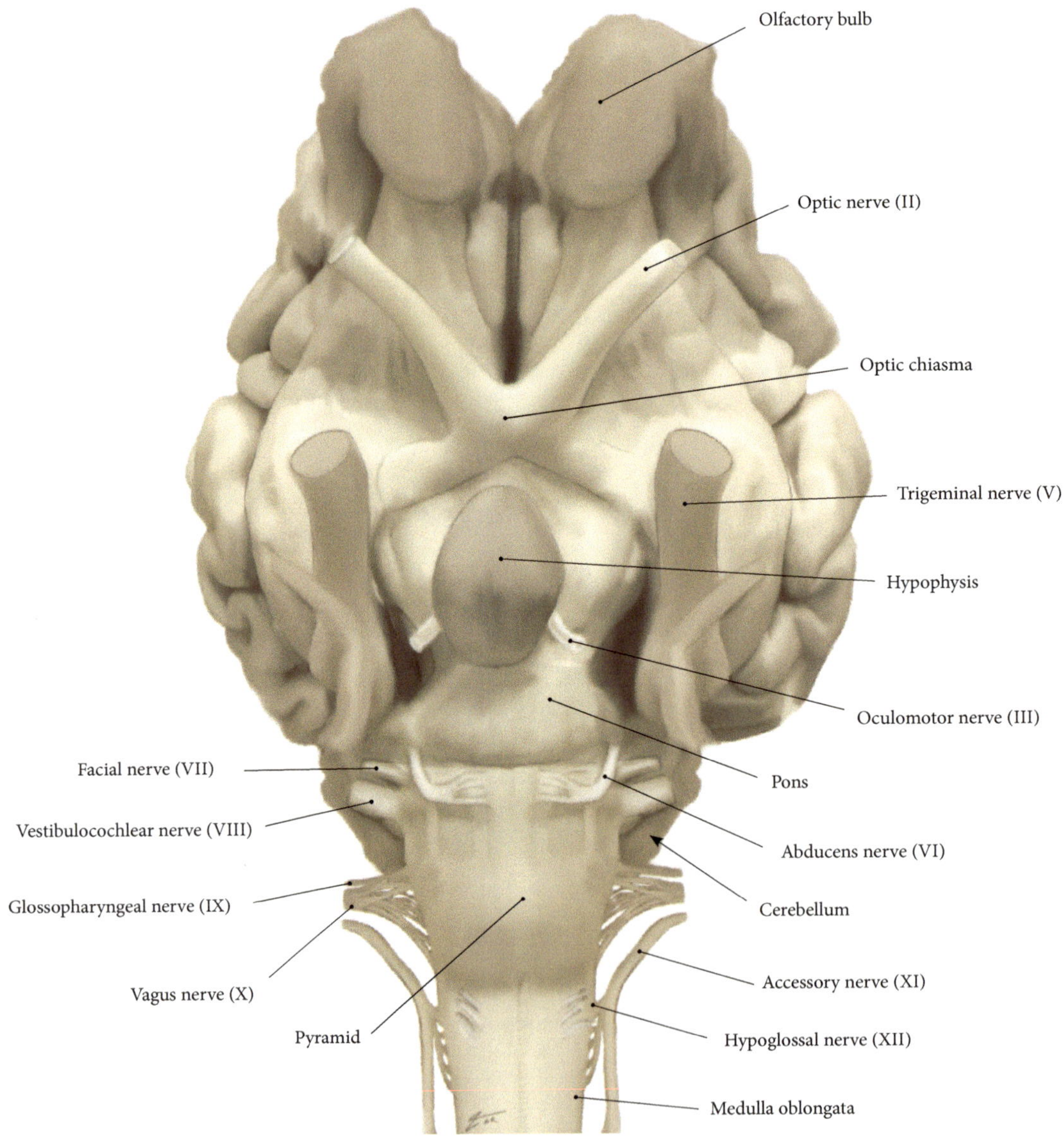

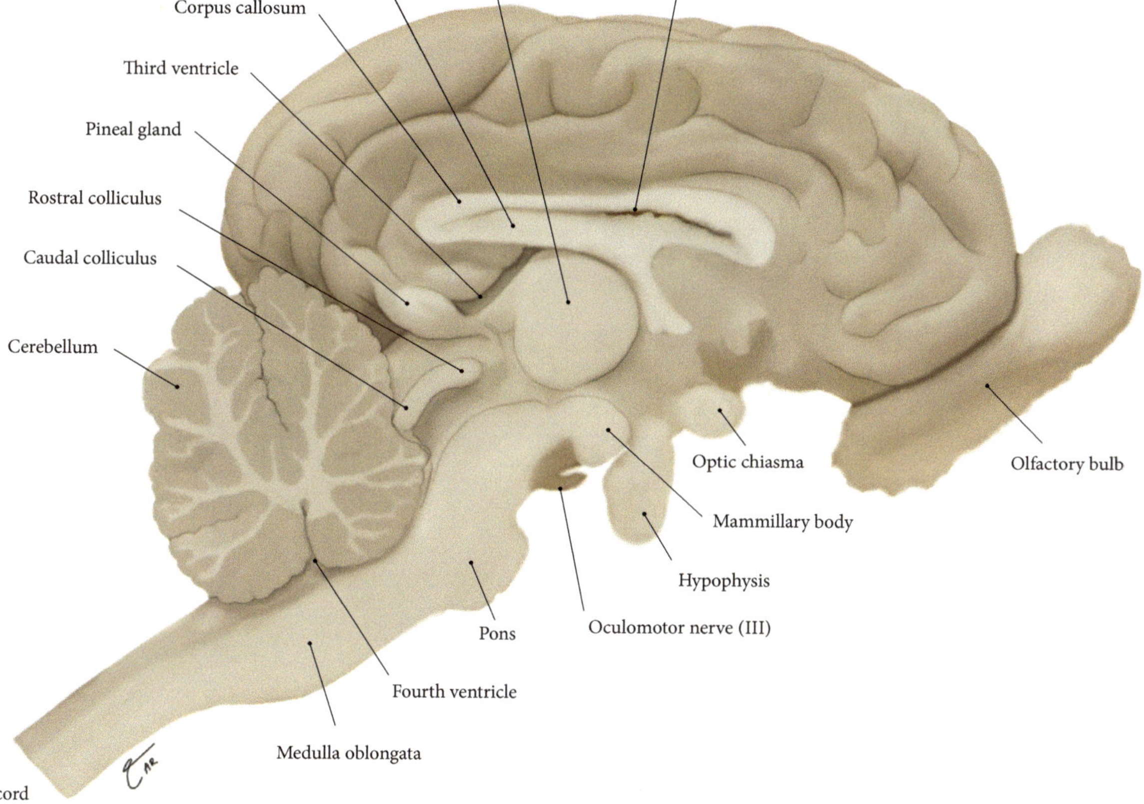
Fornix
Thalamus
Lateral ventricle
Corpus callosum
Third ventricle
Pineal gland
Rostral colliculus
Caudal colliculus
Cerebellum
Optic chiasma
Olfactory bulb
Mammillary body
Hypophysis
Pons
Oculomotor nerve (III)
Fourth ventricle
Medulla oblongata
Spinal cord

Porcine brain – dorsal aspect

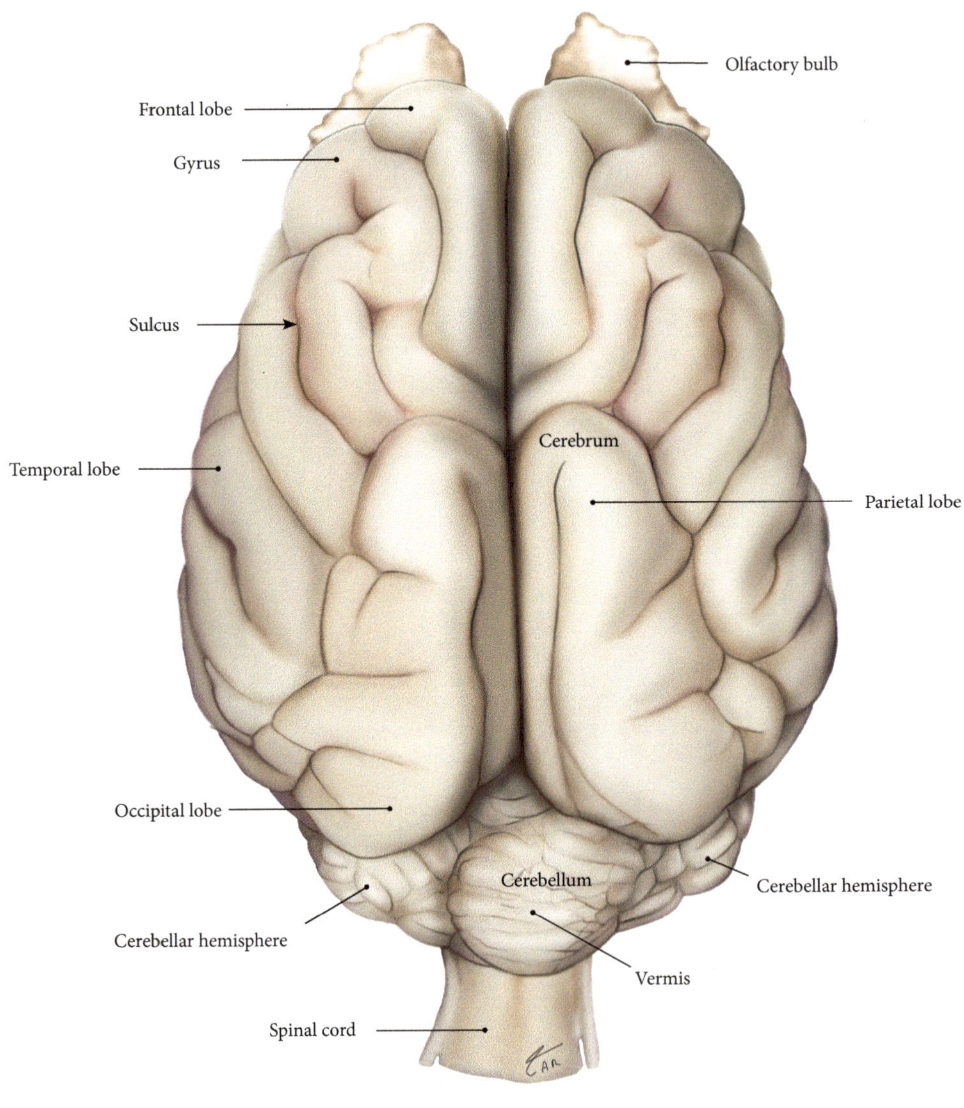

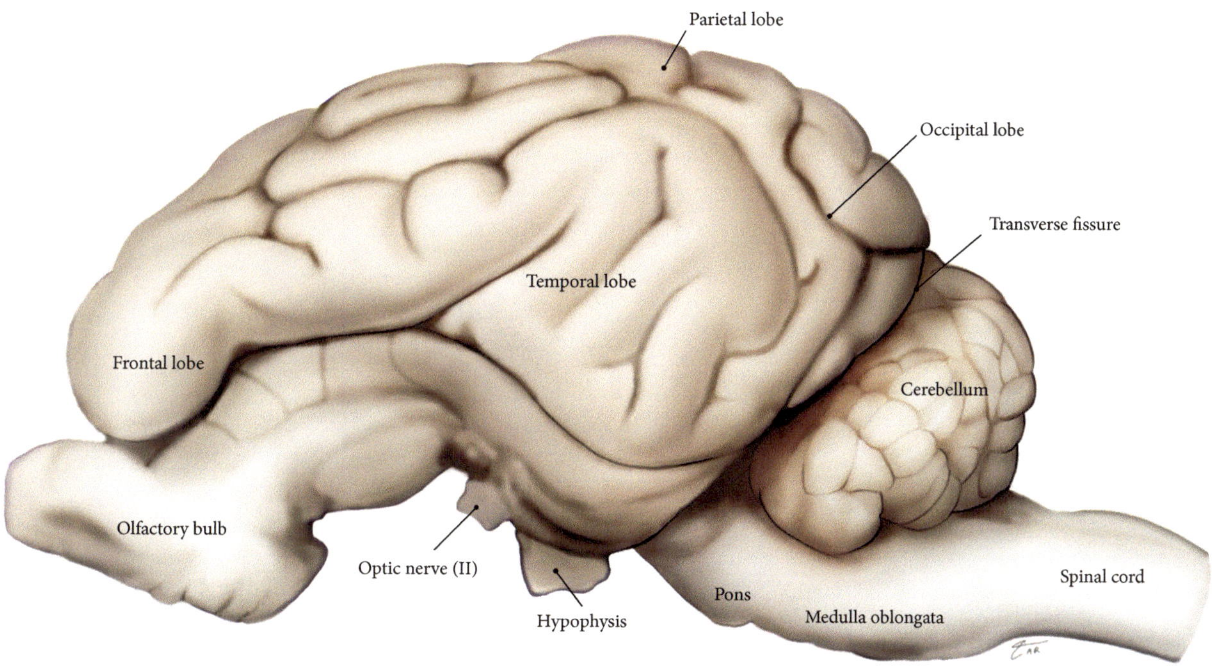
Parietal lobe
Occipital lobe
Transverse fissure
Temporal lobe
Frontal lobe
Cerebellum
Olfactory bulb
Optic nerve (II)
Hypophysis
Pons
Medulla oblongata
Spinal cord

Median sagittal section – epidural anesthesia

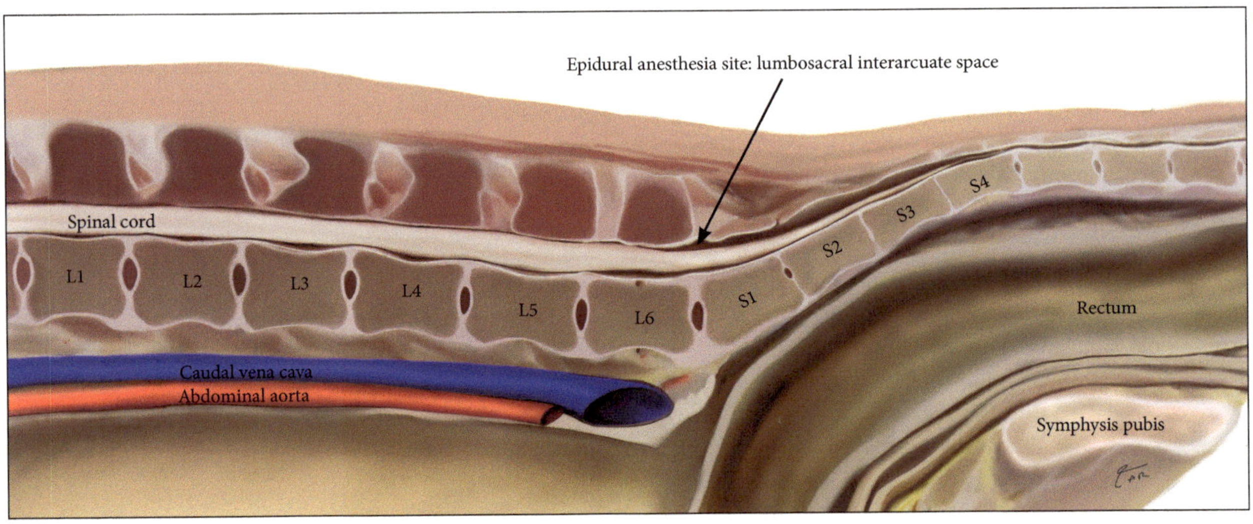

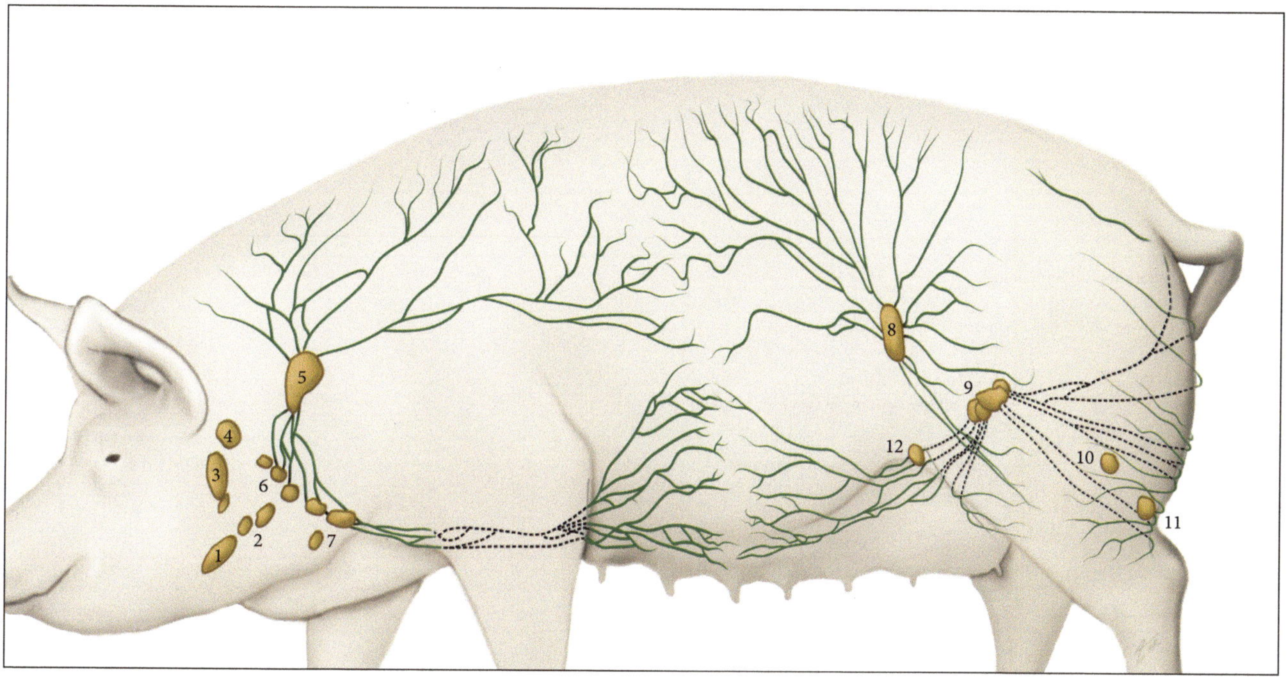

1. Mandibular lymph nodes
2. Accesory mandibular lymph nodes
3. Parotid lymph nodes
4. Retropharyngeal lymph nodes
5. Dorsal superficial cervical lymph nodes
6. Middle superficial cervical lymph nodes
7. Ventral superficial cervical lymph nodes
8. Subiliac lymph nodes
9. Superficial inguinal lymph nodes
10. Deep popliteal lymph nodes
11. Superfical popliteal lymph nodes
12. Accessory mammary lymph nodes

Lymph nodes of the mammary glands

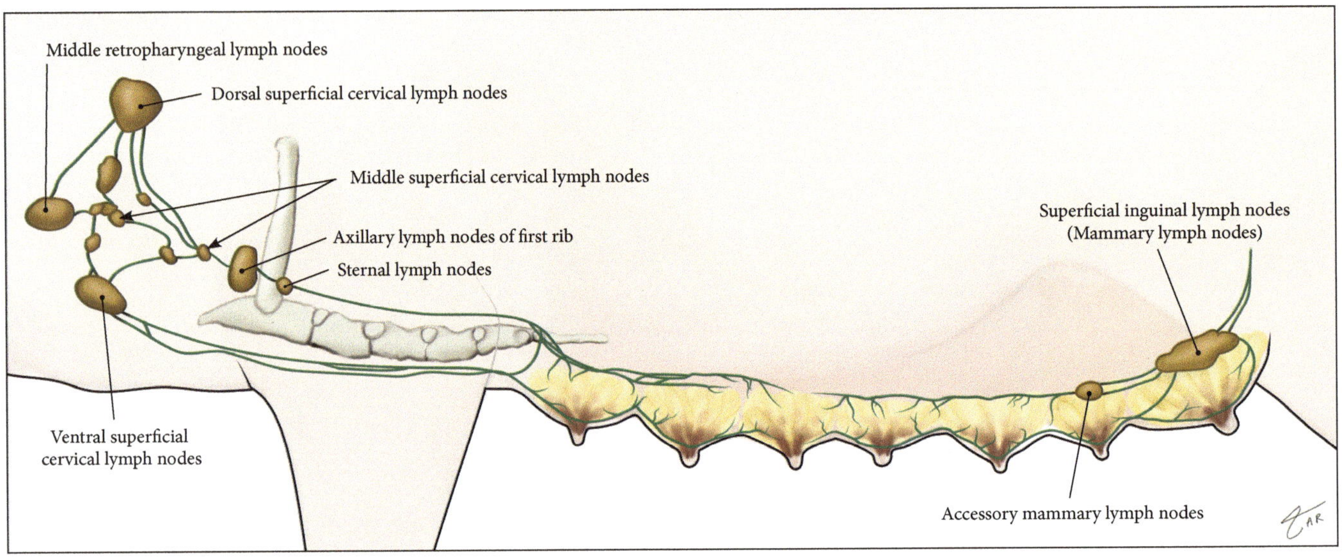

Spleen – visceral surface

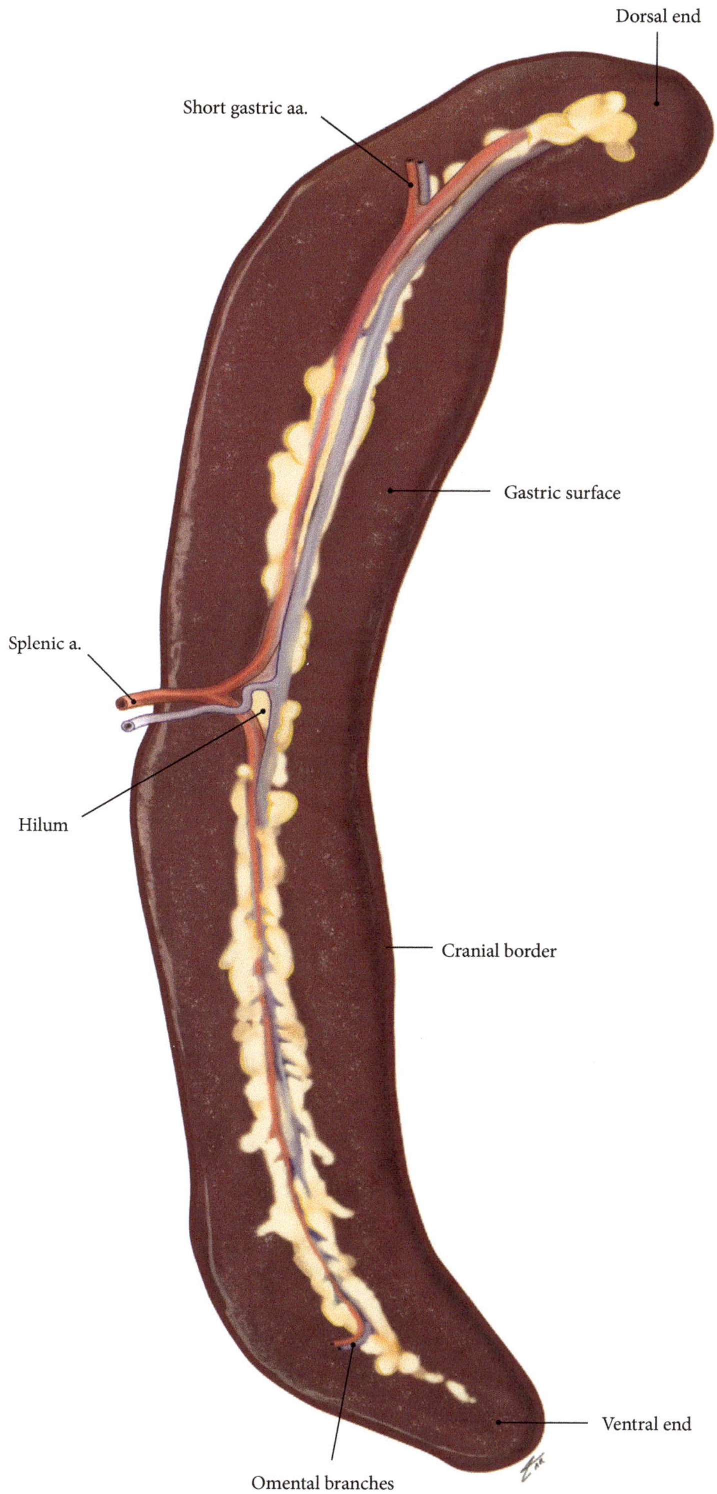

Lymph nodes of the lumbosacral area

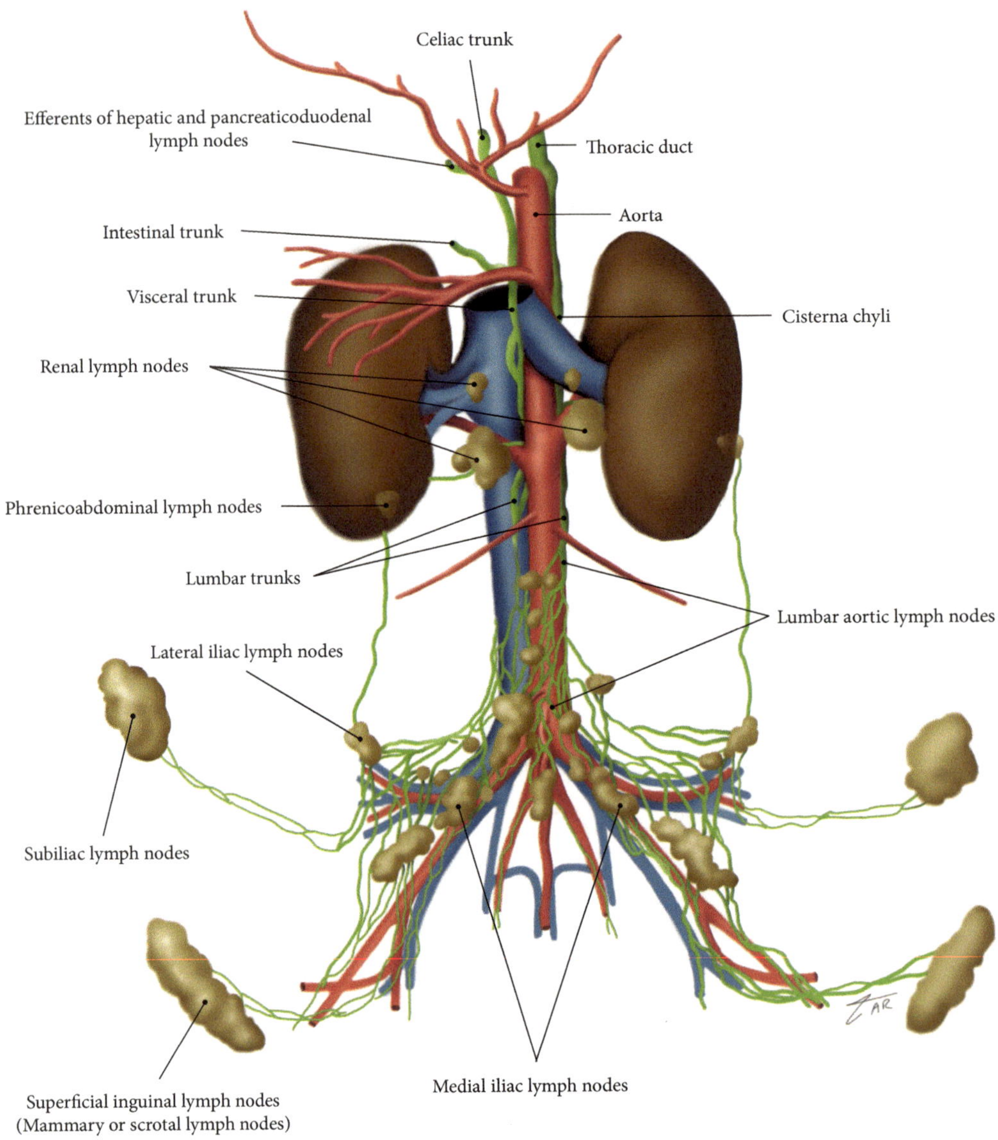

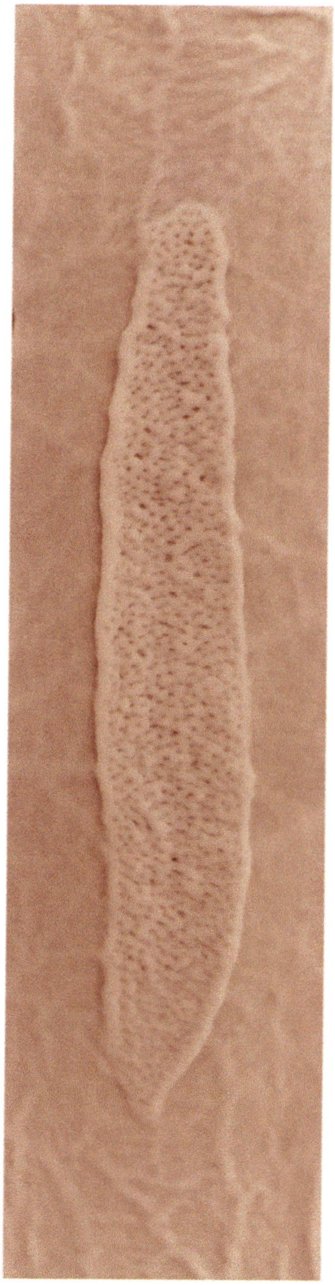

External anatomy

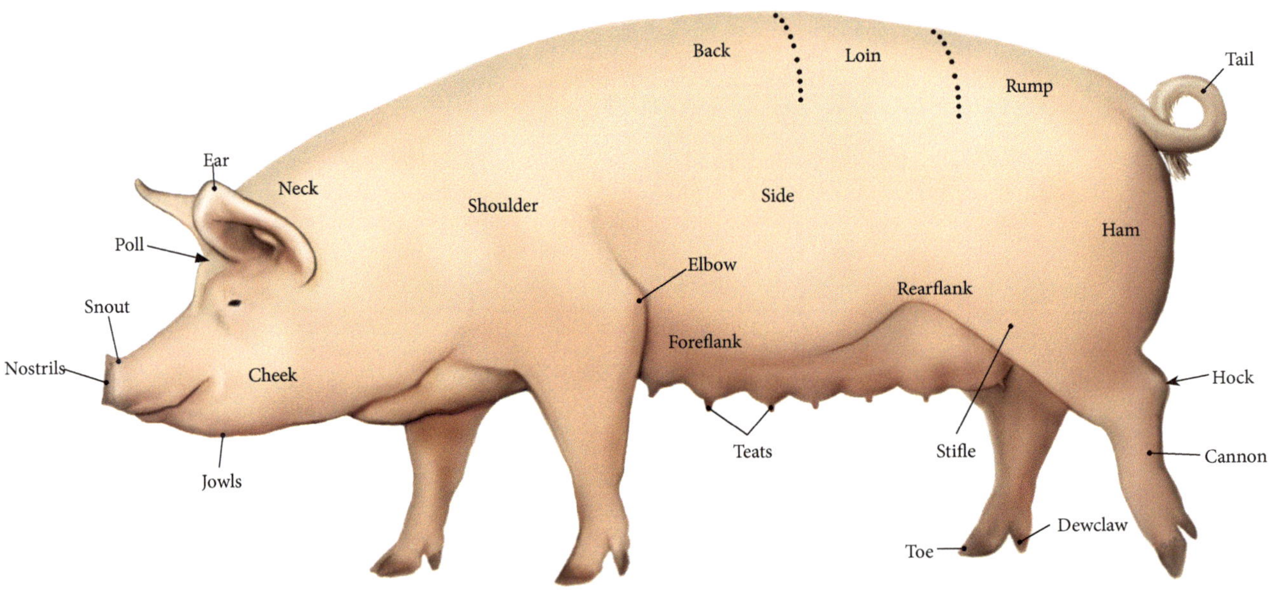

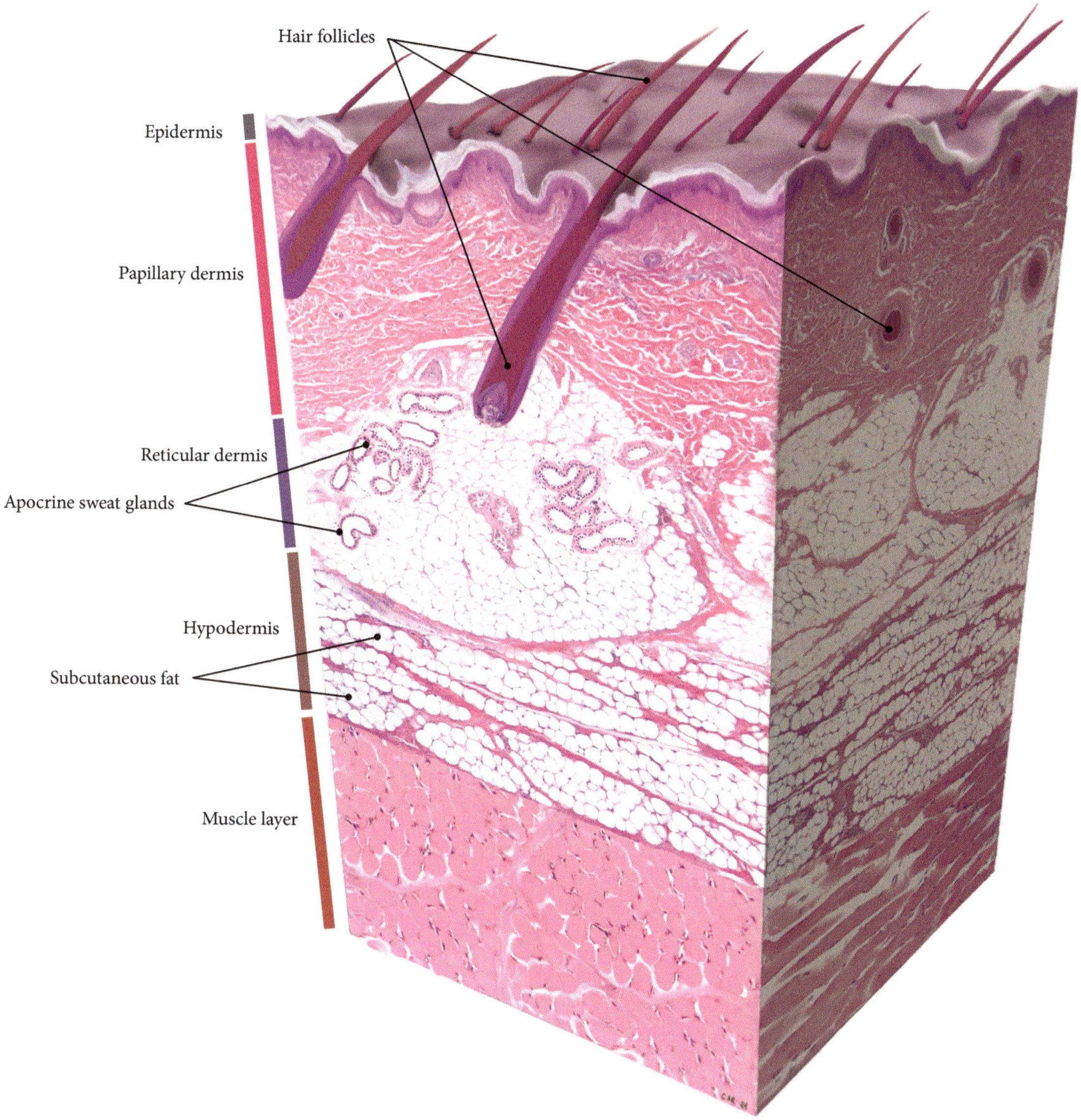
Hair follicles
Epidermis
Papillary dermis
Reticular dermis
Apocrine sweat glands
Hypodermis
Subcutaneous fat
Muscle layer

Mental gland – ventral view

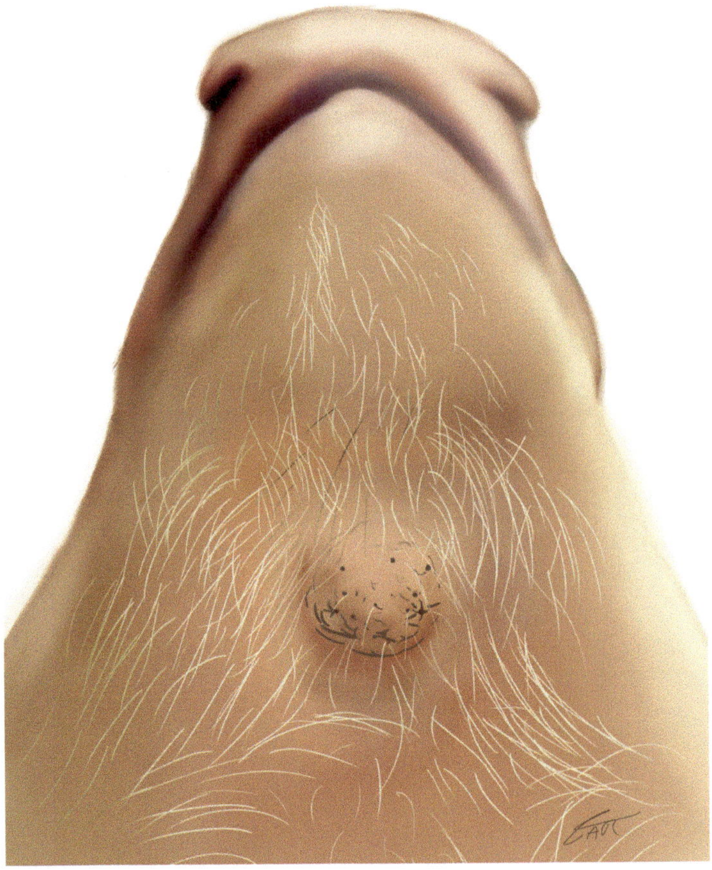

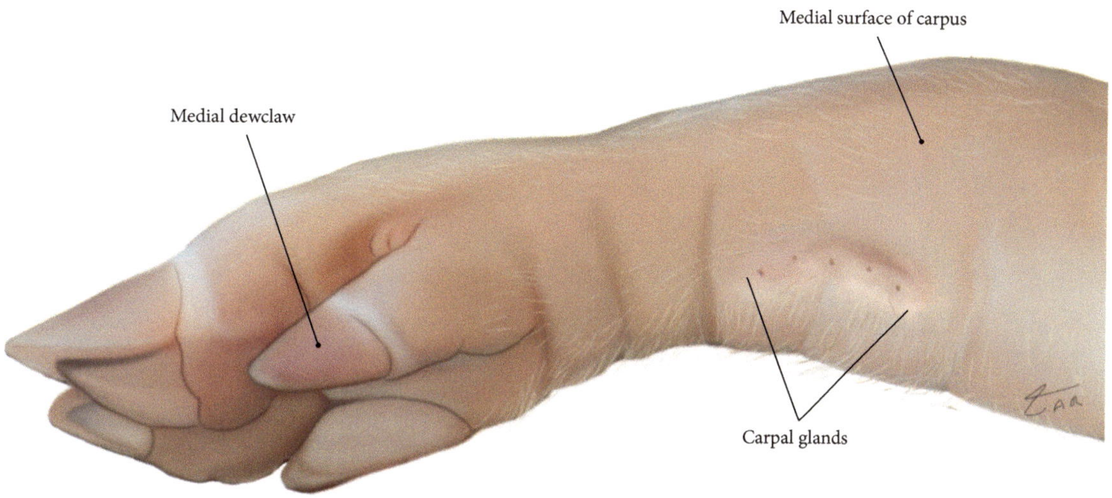
Medial surface of carpus
Medial dewclaw
Carpal glands

Nose and cartilages – craniolateral view

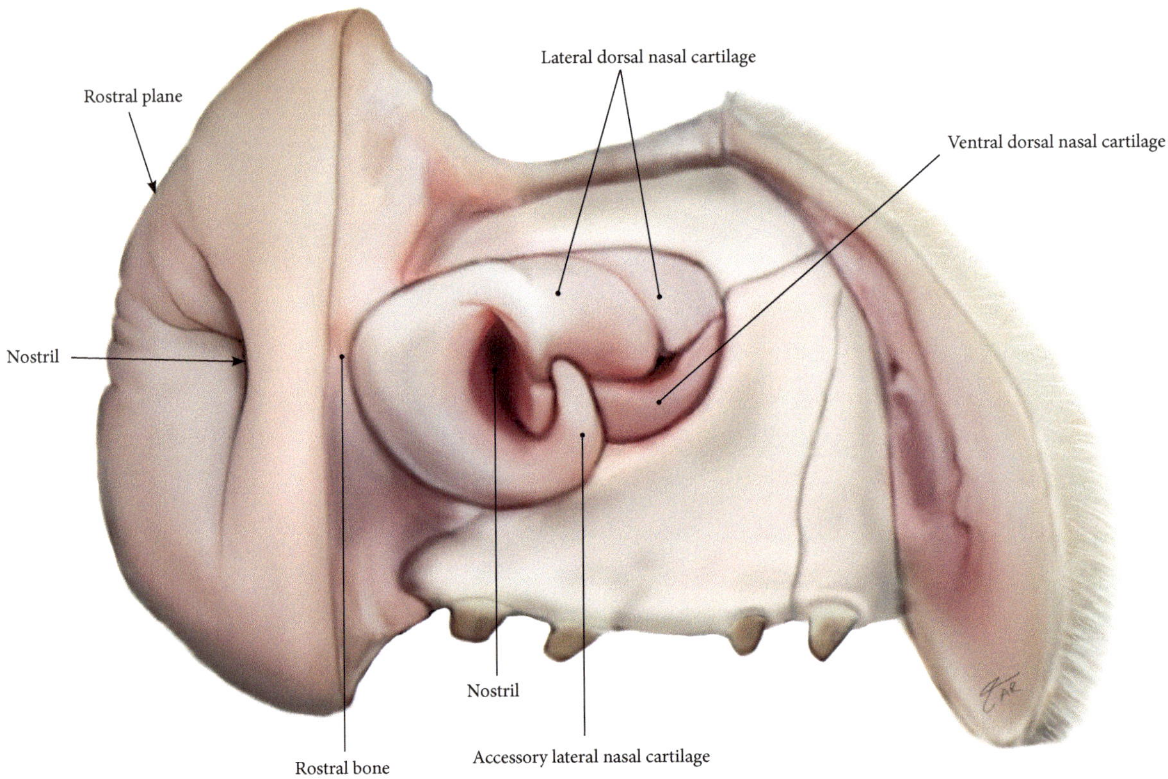

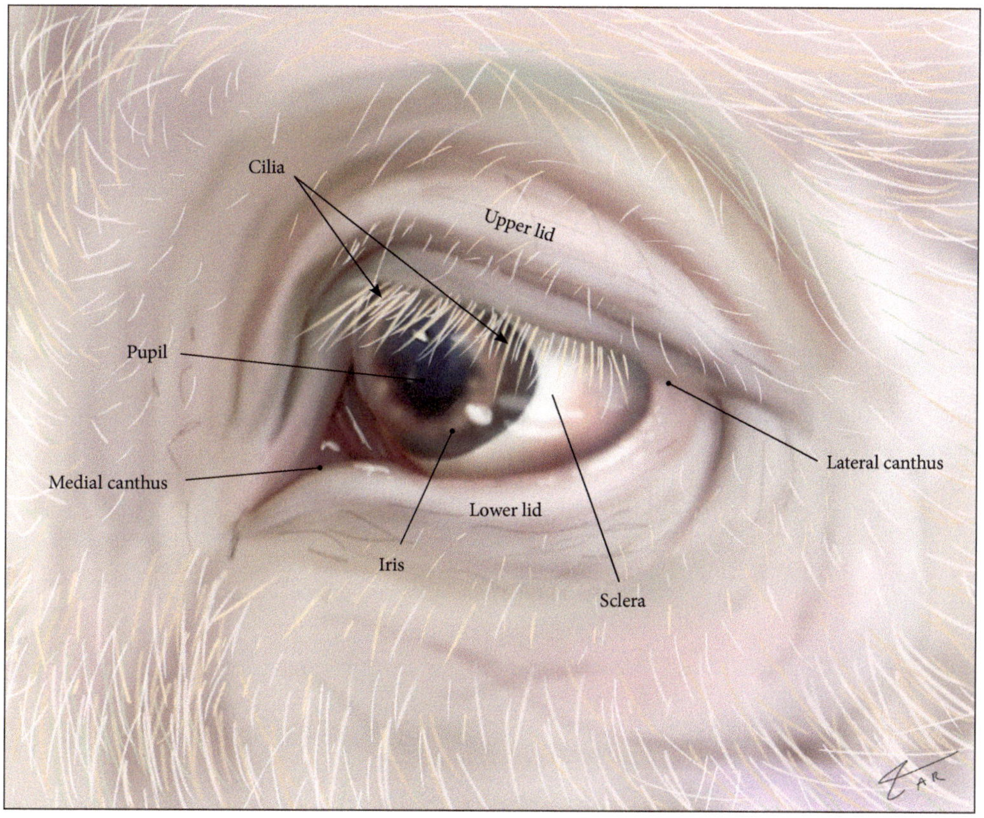
Cilia
Upper lid
Pupil
Lateral canthus
Medial canthus
Lower lid
Iris
Sclera

Eye structures – median section

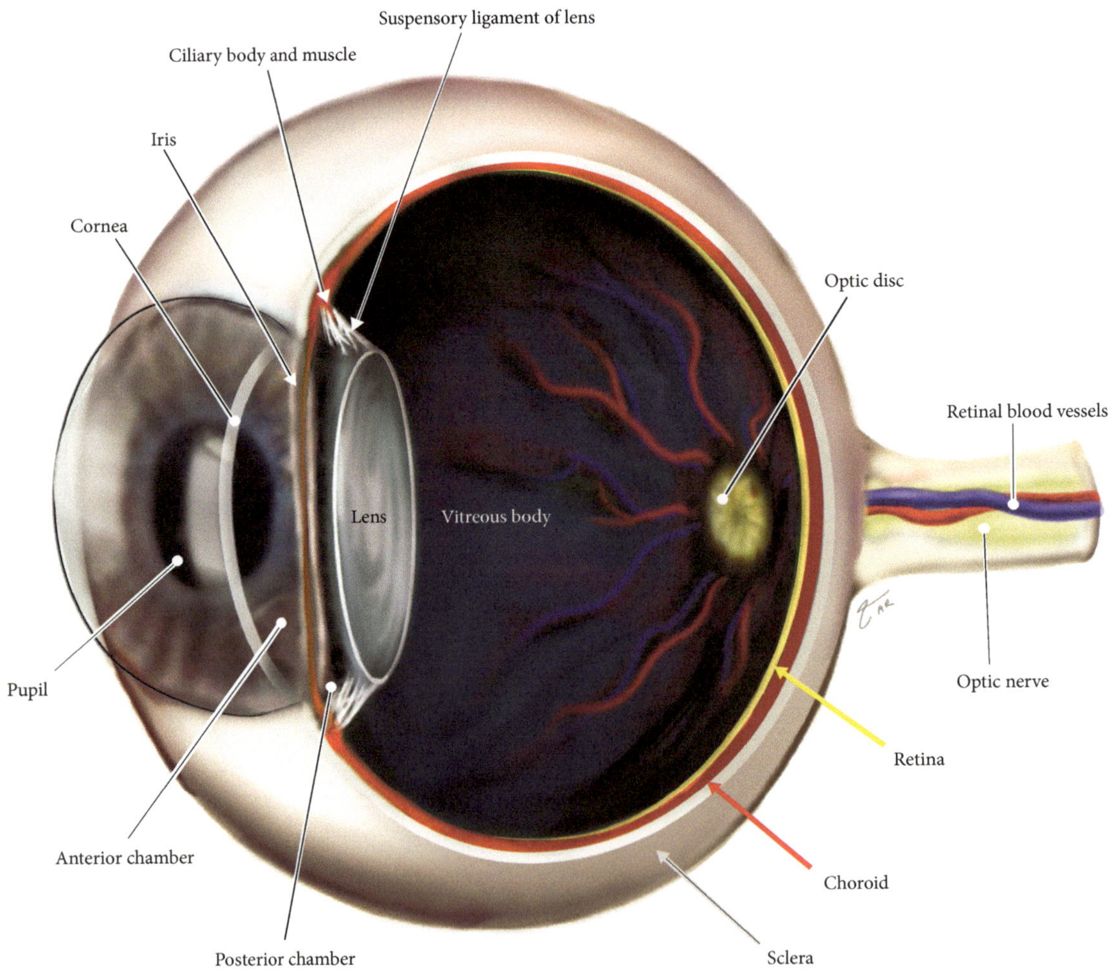

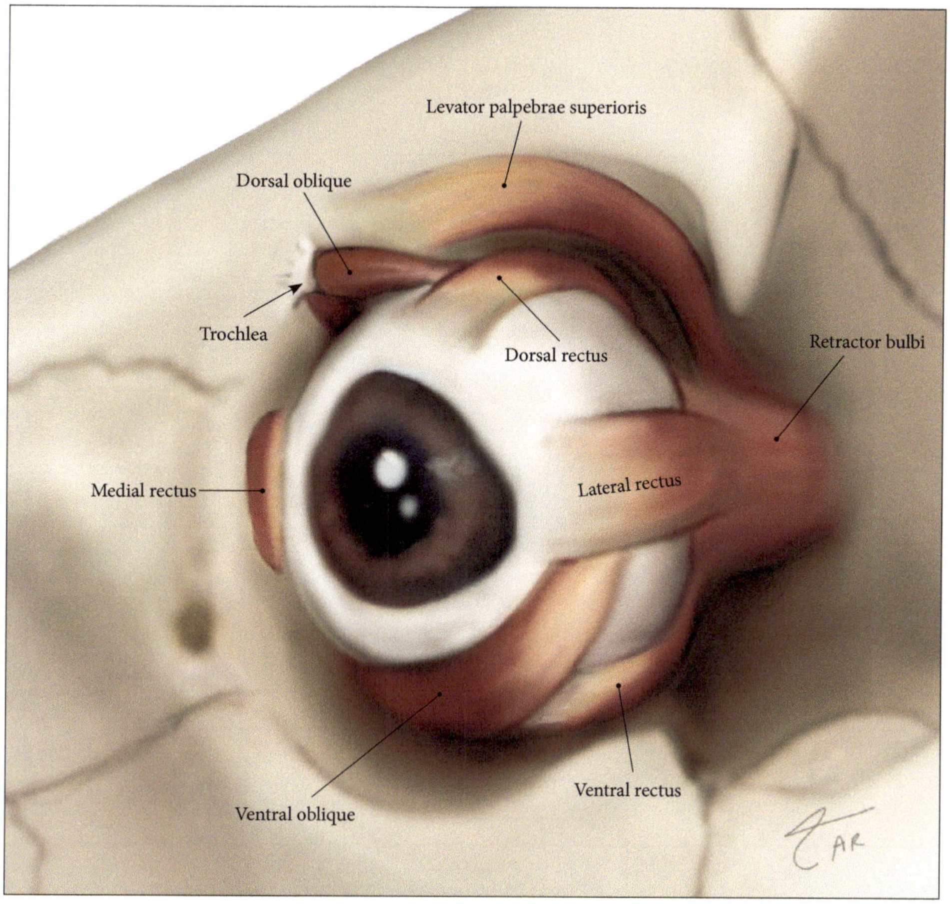
Levator palpebrae superioris
Dorsal oblique
Trochlea
Dorsal rectus
Retractor bulbi
Medial rectus
Lateral rectus
Ventral oblique
Ventral rectus

Median section of the ear canal – rostral aspect

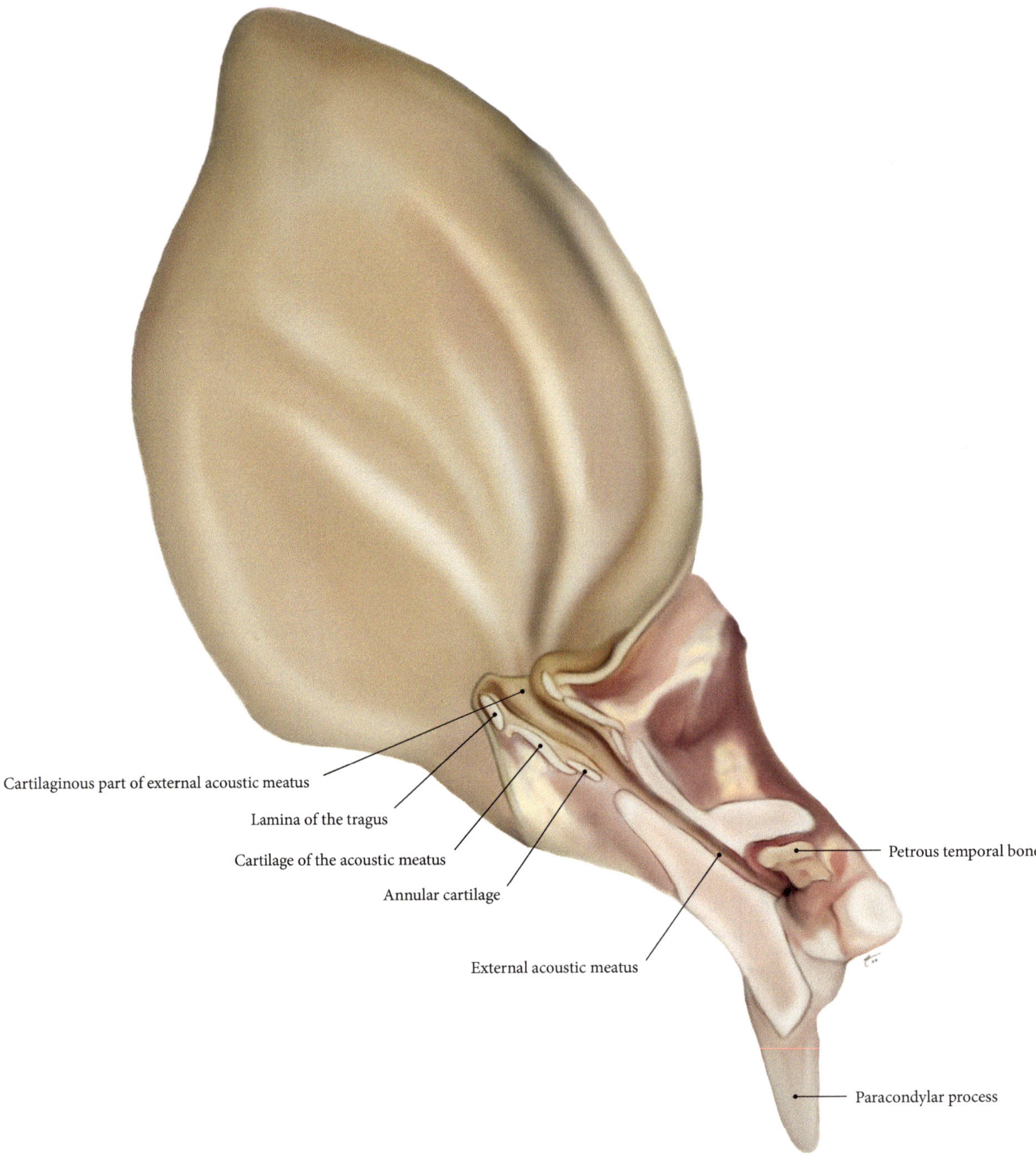

Right ear showing blood vessels

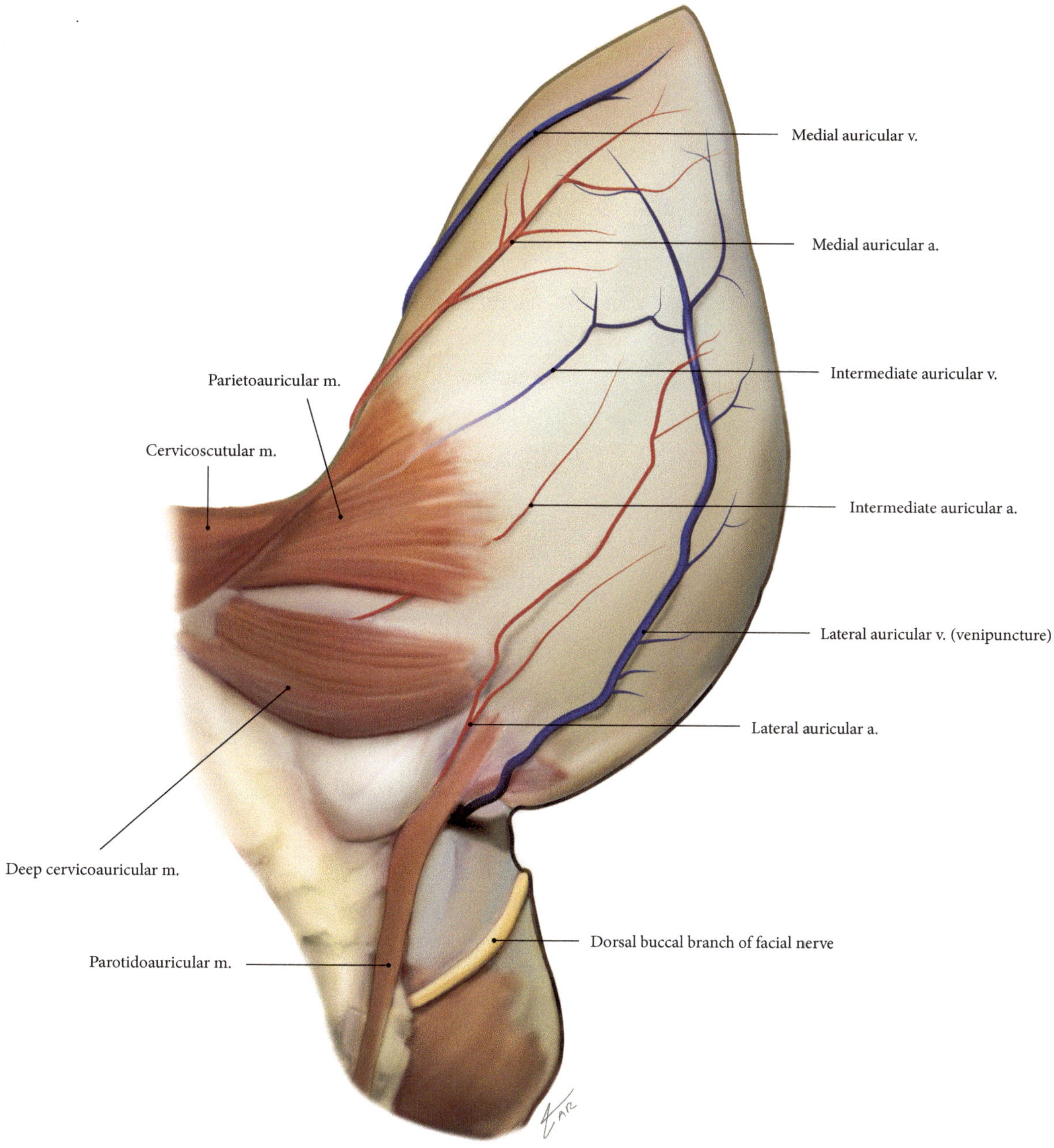

SECTION 3
NORMAL HISTOLOGY OF THE PIG

Kristi Helke, DVM, PhD, DACVP
Professor and Chair, Department of Comparative Medicine
Medical University of South Carolina
Charleston, SC

The reasons for performing a necropsy on a pig are many and varied, and include disease, end of study, and unexpected outcome. Once the necropsy is complete, tissues are commonly examined histologically. In the pig, there are several organs with unique gross and histological features compared to other laboratory animal species. References describing these unique characteristics exist, but for the sake of simplicity, have been collected and summarized in this section.

When performing toxicological or medical device testing studies, sampling should follow guidelines set by the approved authority for the compound or device approval (e.g., USDA, FDA, CE, etc.) The INternational HArmonization of Nomenclature and Diagnostic criteria (INHAND) nomenclature is available and should be used (RENI 2020) in reporting histological findings to regulatory agencies. All of the lesions are indexed on the global open Registry Nomenclature Information system (goRENI) website for regulatory agency-compliant reporting (RENI 2020). This reporting scheme of histological findings has been developed to harmonize reporting among testing sites and is meant to streamline the process of reporting to these agencies.

Fixation of tissues should be performed via perfusion, infusion, or immersion fixation of the tissues with the preferred fixative. Formalin and paraformaldehyde are two commonly used fixatives that provide preservation of tissue architecture as well as good staining characteristics. Davidson's solution is often used for testes to aid in maintenance of morphological structure and prevent artifactual shrinkage of tissues. Tissues sampled should be less than 1 cm^3 if immersion fixation is being used. Tissues should be placed into a container that contains 10× the volume of formalin relative to the weight of the tissue (1:10 w:v ratio).

Pigs can grow to be quite large. Some studies only examine one or few organs; in these cases, methods of infusion fixing isolated organs instead of the whole animal may be used. For the testes, the testicular artery may be used (Lanning et al. 2002). Likewise, the common carotid and external jugular as outflow can be used to infusion fix the brain (Musigazi et al. 2018), and the renal vessels can be used to do an isolated infusion of the porcine kidney (Elling, Hasselager, and Friis 1977).

If immunohistochemistry is going to be conducted, specific fixatives may be required, or special handling of tissues post fixation may be needed (Albl et al. 2016). Immunohistochemistry is commonly used in human and rodent projects, but reagents and porcine-specific antibodies may not be readily available. Unless already published in the pig, most antibodies will need to first be validated in porcine tissue prior to study use.

A complete sampling guide for the pig was published in 2016 that covers different types of studies ranging from routine screening to biobanking (Albl et al. 2016). A few of the gross anatomical peculiarities of the pig compared to other laboratory species will also be mentioned in this section. Common background findings will also be briefly mentioned.

ALIMENTARY TRACT

There are a few anatomical peculiarities in the alimentary tract that are relevant to histology. Occasional foci of chondrocytes may be found in the musculature of the tongue (McInnes and McKeag 2016). In the stomach, the pyloric region contains a fibromuscular protrusion known as the torus pyloricus, the blood vessels of the omentum extend to the small intestinal serosa, and the ascending colon is arranged with centripetal coils (spiral colon) in the abdomen.

When sampling the gastrointestinal tract, it is useful to place the tissues on a stiff flat surface such as a tongue depressor (Figure 3.1) or piece of foil after opening the lumen and gently flushing any ingesta or fecal matter from the surface with phosphate-buffered saline (PBS). The tissue will loosely adhere to the flat surface which can be placed in its entirety into fixative. This technique allows for removal of ingesta including digestive enzymes as well as exposure of the mucosa to fixative. It also allows for more controlled analysis of the organ to be performed as it prevents the tissue from folding in on itself and holds tissue in a uniform conformation.

a b c d

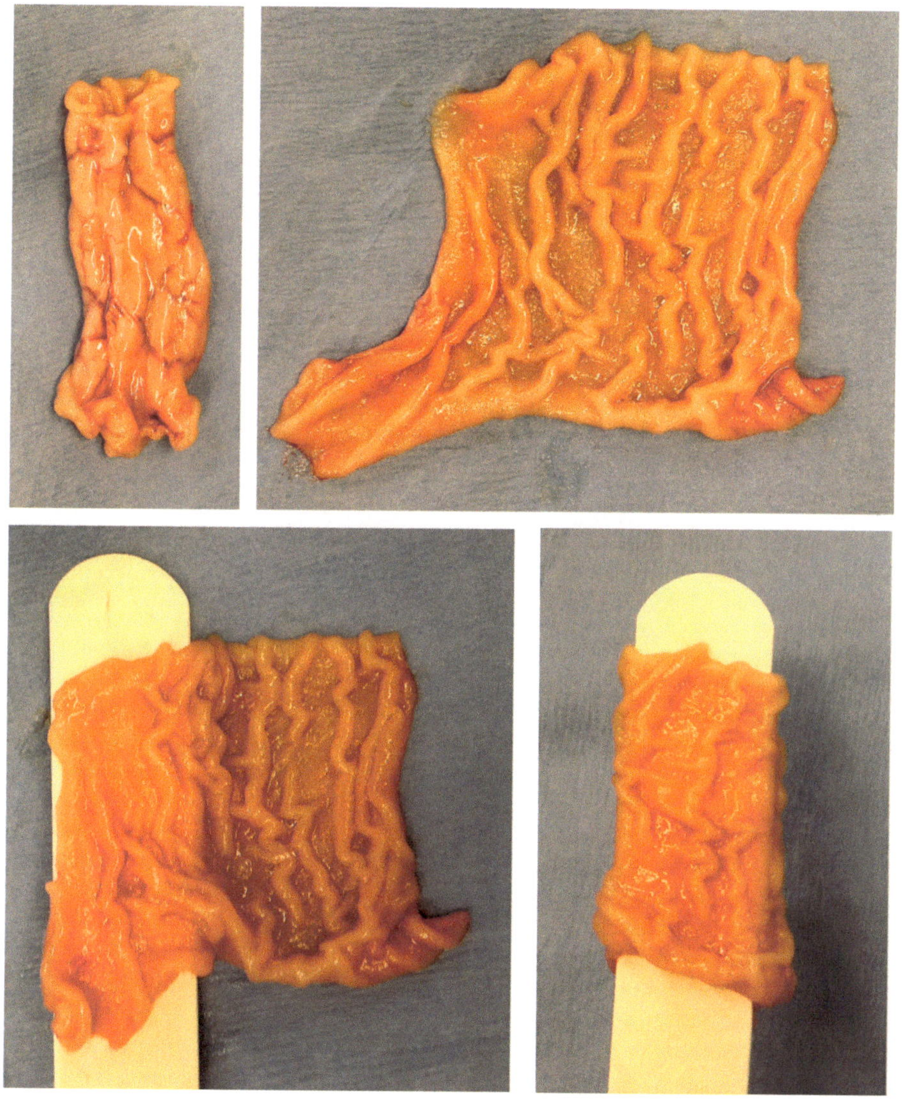

Figure 3.1 a. Intact section of small intestine. **b.** Same section of small intestine opened longitudinally along the amesenteric aspect. **c.** Gentle placement of tissue section on a wooden tongue depressor. **d.** Remaining tissue adhered to opposite side of tongue depressor resulting in flat tissue ready for fixation.

The porcine stomach has four histologically distinct regions. The nonglandular pars esophagea (Figure 3.2) is squamous mucosa which in some instances is keratinized adjacent to where the esophagus enters the stomach. This is also the region which is most likely to show erosion or ulceration. The cardia mainly contains mucus-secreting cells but also has lymphoid nodules within the mucosa (Figure 3.3) (Kararli 1995). Antigenic stimulation is not required for the presence of the lymphoid nodules (mucosa-associated lymphoid tissue [MALT]) as they are present even in young gnotobiotic pigs (Green, Eaton, and Krakowka 1997). The fundus (Figure 3.4) contains parietal and chief cells as does the pyloric region (Figure 3.5a). Within the pyloric region is the torus pyloricus (Figure 3.5b), which is a fibromuscular protrusion at the region of the pylorus that acts as a valve to regulate the release of food from the stomach into the intestine (Bal and Ghoshal 1972).

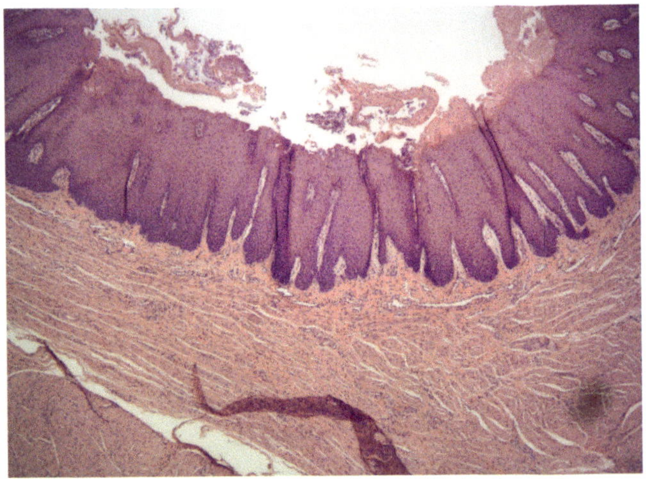

Figure 3.2 Histology of the pars esophagea region of the stomach. Mucosa is stratified squamous with minimal parakeratotic hyperkeratosis. HE 4×.

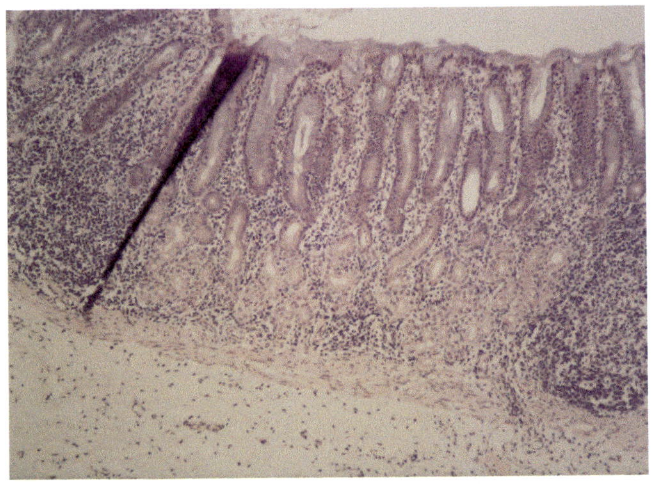

Figure 3.3 Histological section of the cardia region of the porcine stomach. There are abundant mucous cells lining the apical region of the glands. HE 10×.

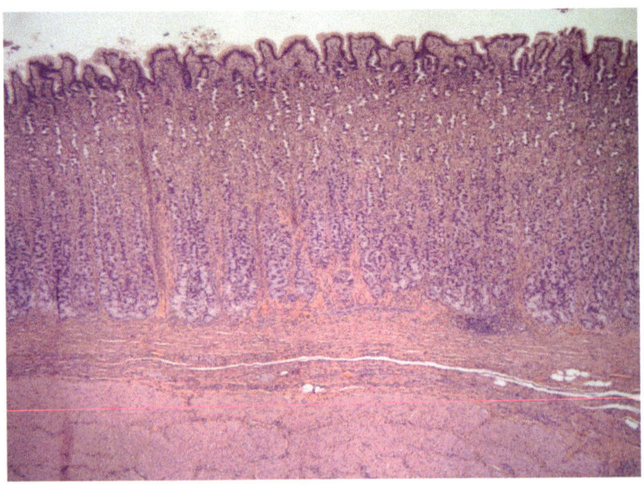

Figure 3.4 Histology of the fundus of the porcine stomach. There are abundant chief and parietal cells within the glands. HE 4×.

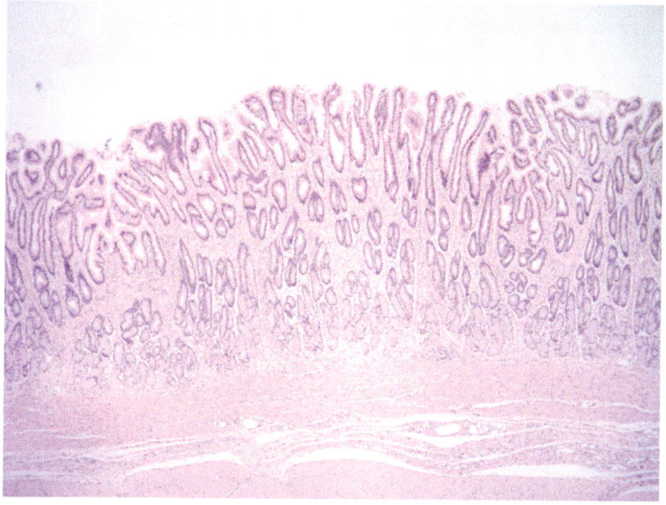

Figure 3.5a Pyloric region of the porcine stomach, mucosa. Parietal cells are abundant and there are few chief cells in this region. HE 4×.

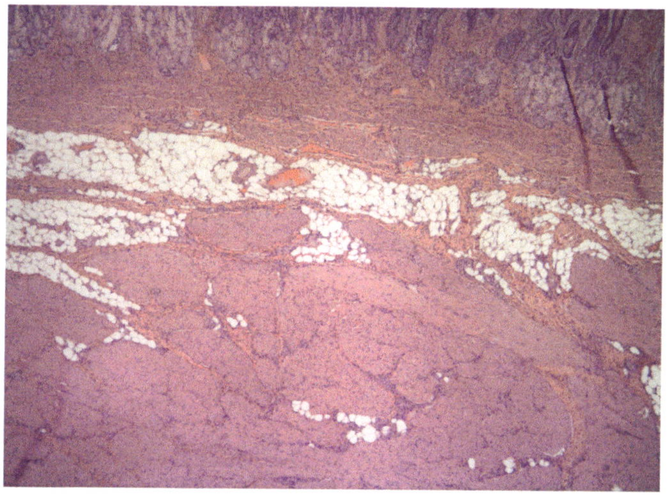

Figure 3.5b Pyloric region of the porcine stomach, torus pyloricus. There are thick muscular bundles arranged in varying directions within the torus pyloricus. HE 4×.

The small intestine consists of the duodenum, jejunum, and ileum (Figures 3.6–3.8a). Similarly to humans, pigs have plicae circularis covered with villi which contain enterocytes with microvilli at the apical surface, resulting in increased absorptive surface area. The length of the porcine small intestinal villi relative to rodent/human villi is similar. Both mucous goblet cells and to a lesser extent enteroendocrine cells increase in number distally within the small intestine (Gonzalez, Moeser, and Blikslager 2015). There is controversy as to whether or not pigs have Paneth cells in the crypts of the small intestine (Gonzalez, Moeser, and Blikslager 2015). A cell with similar structure has been identified, but it lacks certain features such as the distinct Paneth cell granules and lysozyme has not yet been detected in these cells (Gonzalez et al. 2013).

Pigs have two types of gastrointestinal-associated lymphoid tissue (GALT): discrete Peyer's patches in the wall of the jejunum and upper ileum, as well as a continuous area of lymphoid tissue in the distal ileum. The amount of GALT increases with age in the pig (Figure 3.8b) (Pabst et al. 1988).

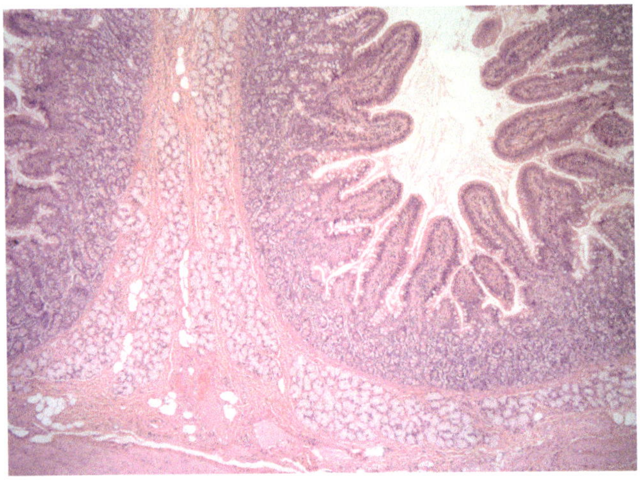

Figure 3.6 Duodenum. The mucosa is arranged in villi overlying plicae which contain abundant Brunner's glands. HE 4×.

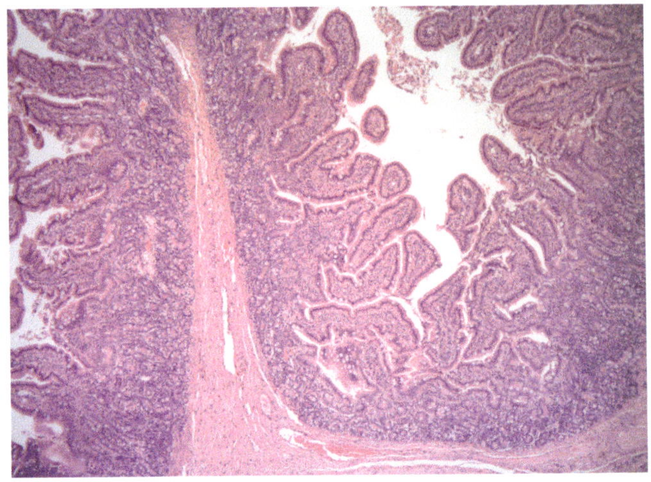

Figure 3.7 Jejunum. Villi with occasional mucous goblet cells overlie plicae. HE 4×.

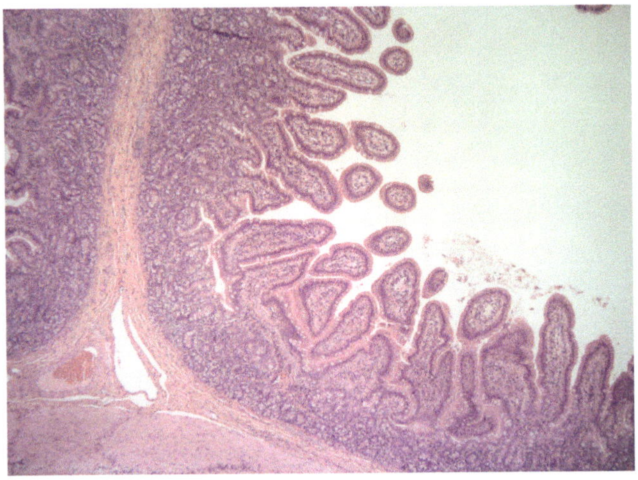

Figure 3.8a The porcine ileum contains abundant villi with scattered mucous goblet cells arranged over plicae. HE 4×.

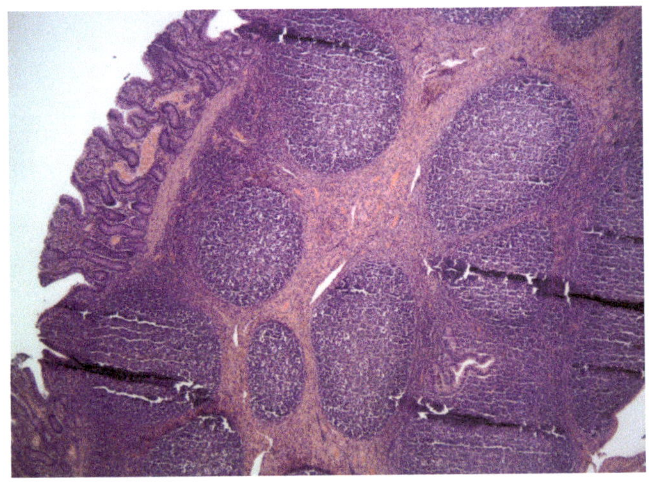

Figure 3.8b Within the muscularis layer of the ileum is an area with abundant lymphoid tissue that is visible grossly. Histologically there are numerous lymphoid follicles. HE 4×.

The ascending colon is arranged in centripetal coils and is referred to as the spiral colon (Yen 2001). Unlike rodents, in the pig, all sections of the colon (Figure 3.9) contain taeniae coli, longitudinal smooth muscle bands along the colonic serosa that produce the haustra or sacculations of the colon (Hedemann, Kristiansen, and Brunsgaard 2002). The histology of the porcine colon is similar to that of the human, with abundant mucous cells.

The alimentary tract is very plastic and all gastrointestinal parameters are dependent on numerous factors such as sampling site as well as diet. Care must be taken to ensure tissues are sampled from the same anatomical location in all animals of a study. This applies not only to the mucosa, but also to the taenia and haustra, as they also differ based on anatomical location (Hedemann, Kristiansen, and Brunsgaard 2002). Histology of villi differs based on feed components (Mekbungwan, Thongwittaya, and Yamauchi 2004) and villous length may increase in the ileum with increased social stress (Li et al. 2017).

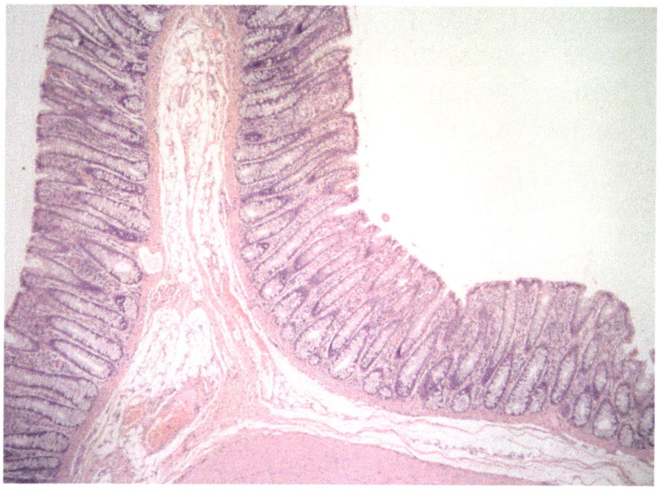

Figure 3.9 The mucosa of the spiral colon has numerous mucous goblet cells lining the crypts which overlie plicae. HE 4×.

CARDIOVASCULAR SYSTEM

The porcine heart is similar in many ways to the human heart. The histology is also quite similar between the two species. One difference is that the Purkinje fibers are quite prominent in pigs (Figure 3.10), but not visible in the human subendocardial region (Sedmera and Gourdie 2014). These fibers are not of nervous, but myogenic origin and have striations (Sedmera and Gourdie 2014). Telocytes are a recently described interstitial cell type that has been described in many organs including the left and right atria and ventricles (Tay, Vandecasteele, and Van den Broeck 2017). These cells have a small cell body and are believed to have a role in intercellular communication (Tay, Vandecasteele, and Van den Broeck 2017). Vessels in pigs have a thicker adventitia and smooth muscle layer as well as increased density of elastic fibers compared to other species (Simões et al. 2016). Similarly to other large mammals, the large vessels of the pig contain a network of small vessels, the vasa vasorum, within the media and adventitia to provide nutrients to the vascular wall (Figure 3.11) (Mulligan-Kehoe 2010).

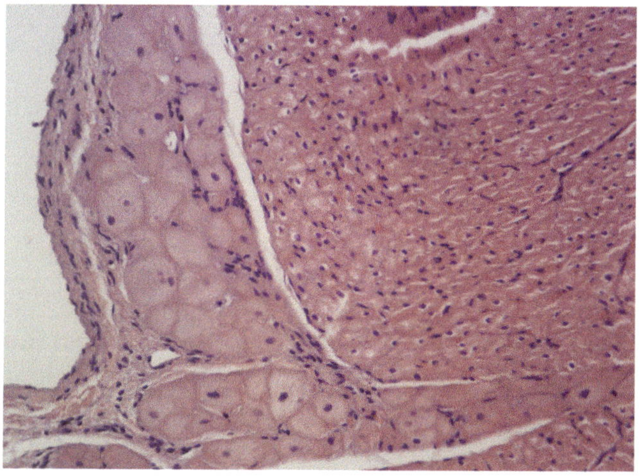

Figure 3.10 There are modified cardiomyocytes on the left which are larger in diameter just under the endocardium. These Purkinje cells are plump and paler in staining than the adjacent cardiomyocytes. HE 20×.

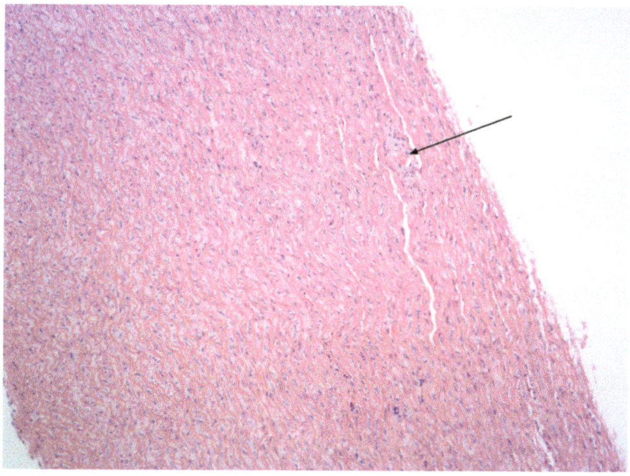

Figure 3.11 Aorta. There are vessels of varying sizes within the wall of the aorta and other large vessels in the pig. HE 10×.

Arteritis/periarteritis may occur spontaneously in Göttingen minipigs, most commonly in cardiac and extracardiac blood vessels, but also in other organs. These cellular accumulations are primarily composed of mixed inflammatory cells, both within and around blood vessels, and in some cases may damage the vascular wall and endothelium and occlude the lumen (Dincer et al. 2018).

ENDOCRINE SYSTEM

Adrenal Glands

The adrenal gland is composed of a cortex and medulla. All three layers of the adrenal cortex (zona glomerulosa, zona fasciculata, and zona reticularis) are evident in the pig (Figure 3.12). Cortical (17-alpha-hydroxylase-positive) cells are commonly found within the medulla (Bornstein et al. 1991), and rays of chromaffin-positive (medulla) cells are found within the cortex. There is often a layer of lipid-laden cells separating the cortex from the medulla. This layer remains in the same position in males as they age, whereas in females, it is found within the zona reticularis as the animals age (Fujioka 1956). In females, after 2 years of age, the zona reticularis will comprise over 50% of the cortex, whereas it is not observed in males of this age (Fujioka 1956).

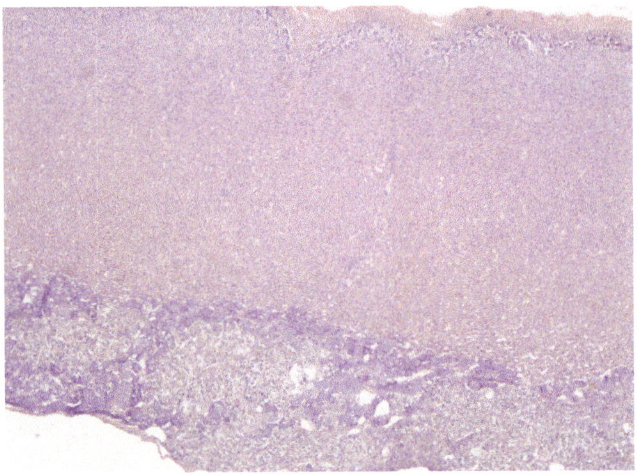

Figure 3.12 The porcine adrenal gland has a discernable zona glomerulosa in the cortex. The medulla has two distinct populations of cells (adrenergic and noradrenergic) which are arranged in distinct islands and not intermixed. HE 4×.

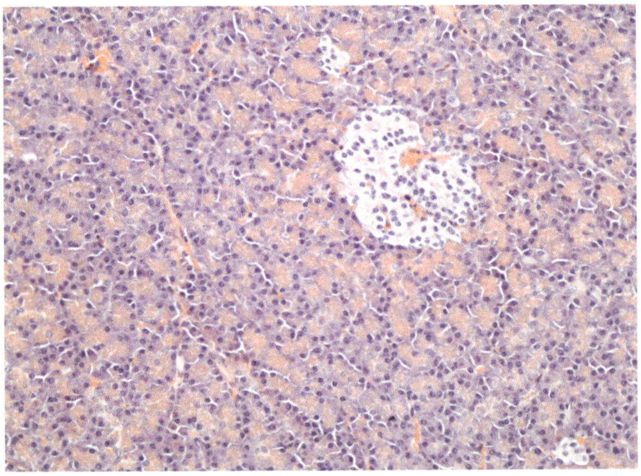

Figure 3.13 The porcine pancreas is histologically similar to the human pancreas. There are islets composed of neuroendocrine cells surrounded by the exocrine pancreas. HE 20×.

Pancreas

In the pig, the endocrine pancreas comprises less than 2% of the total organ. The endocrine cells (alpha, beta, glucagon cells) in the islets of the pig are intermingled with no geographical predilection for the cell types (Tsuchitani, Sato, and Kokoshima 2016); otherwise, the pancreas is histologically similar to that of other species (Figure 3.13).

Parathyroid Glands

The pig only has one pair of parathyroid glands that are located within the cervical portion of the thymus near the bifurcation of the carotid artery into internal and external carotid arteries (Oster et al. 2018; Soshin et al. 2010). The thymus may encircle the parathyroid, or thyroid tissue may be found adherent to or within the thin parathyroid capsule (Figure 3.14a). The histology of the parathyroid consists of a thin fibrous capsule and septae separating chief cells (Figure 3.14b). Adipose cells may be found in the parathyroid gland of older animas. Thymic tissue may infiltrate the parathyroid gland and be seen in tissue sections (Oster et al. 2018).

a

b

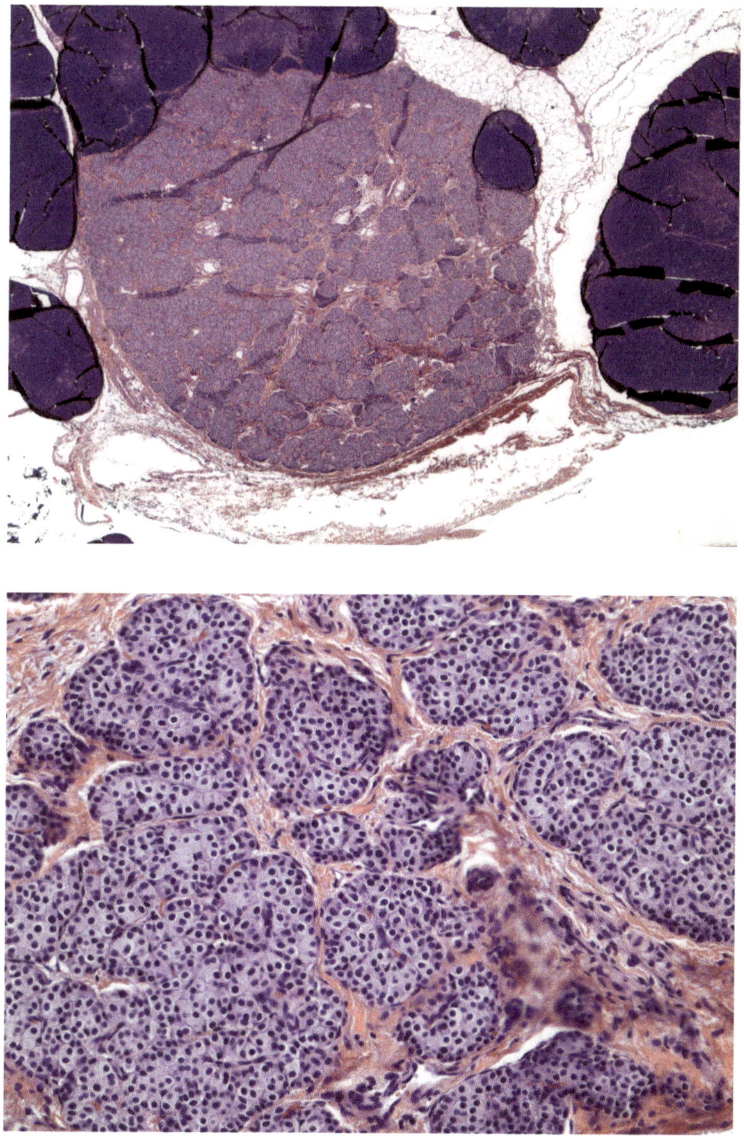

Figure 3.14 a. The porcine parathyroid gland is embedded within and closely opposed to the cervical thymus tissue. HE 2×. **b.** The parathyroid gland in the pig is histologically similar to that of other species, with nests of chief cells. HE 20×.

Thyroid Gland

The porcine thyroid gland is bilobed, but the lobes are fused giving the appearance grossly of a single uniform gland on the ventral surface of the trachea. Thyroid follicles, follicular number, and capsule thickness increase with maturation in the pig (Figure 3.15) (Igbokwe and Ezeasor 2015). Hemorrhage is commonly seen in the thyroid gland and surrounding tissues due to blood sampling (Dincer and Skydsgaard 2012; Rinke 1997). If blood sampling is frequent, fibrosis may also be present (Helke et al. 2016).

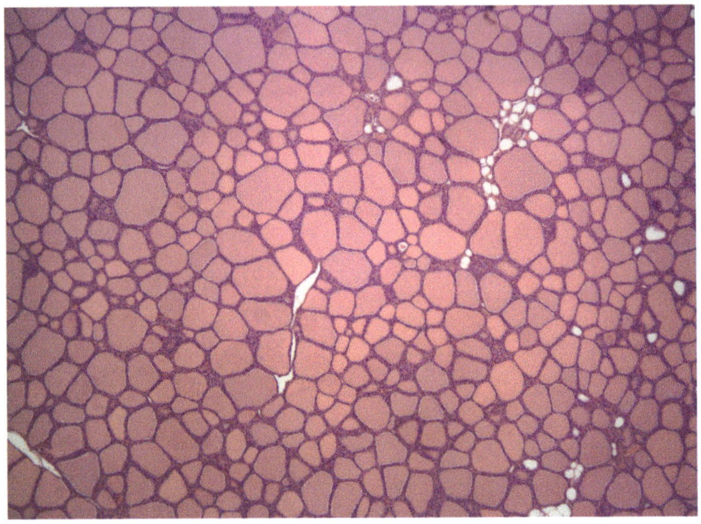

Figure 3.15 The thyroid gland of the pig is similar to that of other species. HE 4×.

HEMATOLYMPHOID SYSTEM

Hemal Nodes

Hemal nodes are common in ruminants, but a few recent reports describe them in the pig as well. Hemal nodes have been noted in the thoracic, abdominal, and pelvic regions of the pig along blood vessels (Udoumoh, Ezeasor, and Igwebuike 2016). They are connected with blood vessels and covered with a thin connective tissue capsule. No lymph vessels are observed within the hemal nodes, although lymphocyte infiltration may be observed (Akaydin and Kabak 2006). Capillaries, smooth muscle cells, fibroblasts, and reticular cells are present within the thin capsule of the hemal node (Akaydin and Kabak 2006).

Lymph Nodes

In pigs, the entire structure of the lymph node is reversed compared to what is typical in other species, including the structure of the lymph follicles, with the cap of lymphocytes on the follicle being on the deep aspect whereas in other species, it is oriented outwardly towards the capsule (Figure 3.16) (Hunt 1968). In pigs the lymph flow is also reversed, as it enters the node centrally and exits peripherally; thus there are no true medullary cords in the pig (Haley 2017). In the mesenteric lymph node, the arteries into the lymph node subdivide into branches and form a plexus of vessels enveloping the lymph node, known as the perimesenteric plexus, which then enters the parenchyma (McInnes and McKeag 2016).

In all other lymphoid tissue in the pig, such as MALT and Peyer's patches, the germinal centers are oriented towards the surface, as is seen in other species.

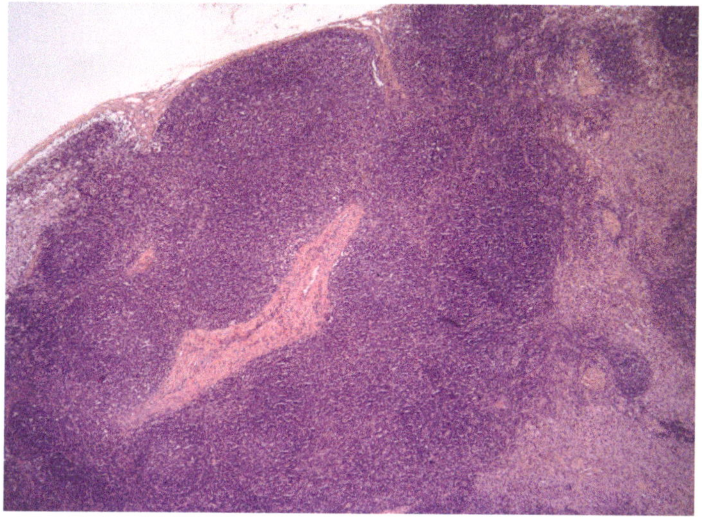

Figure 3.16 The anatomy of the porcine lymph node is reversed from most other species. The follicles are found in the medulla, with lymph flow also reversed. HE 4×.

Spleen

Pigs have an intermediate-type spleen which is a mix of a defensive and storage spleen (Haley 2017). This means that there is an intermediate amount of smooth muscle with an intermediate capability for contraction and thus storage of blood within the spleen (Figure 3.17a) (Boes and Durham 2016). The splenic capsule is composed of smooth muscle, collagen, and elastic fibers (Shringi et al. 2018). Within the parenchyma, there are prominent accumulations of phagocytic cells surrounding capillaries, which are normal in the pig (McInnes and McKeag 2016) and are commonly referred to as Schweigger-Seidel sheaths or ellipsoids (Figure 3.17b).

a

b

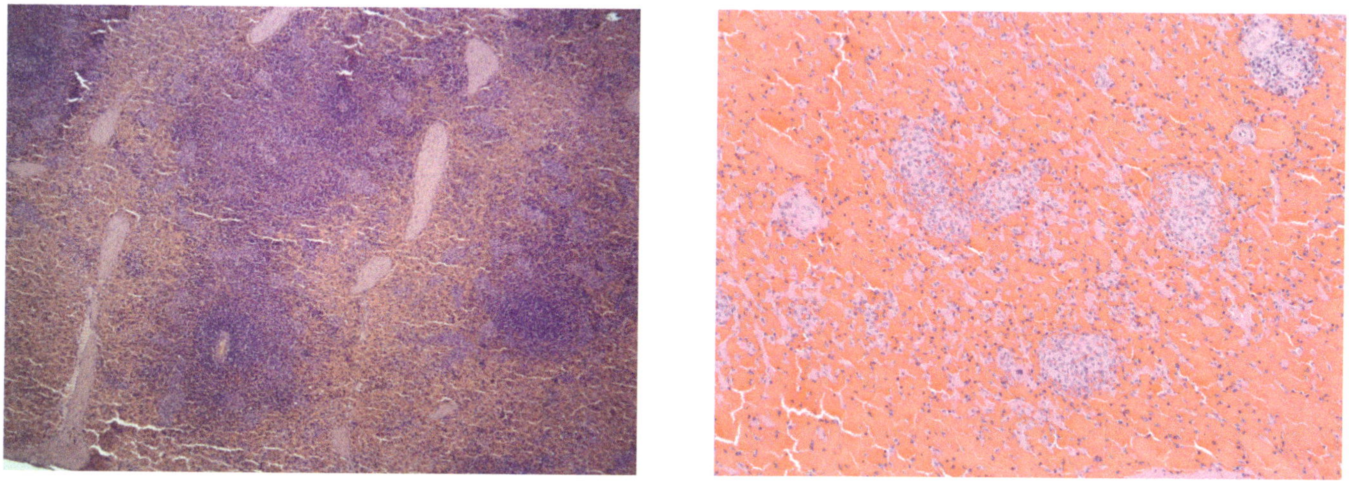

Figure 3.17 a. The spleen has follicles, and the red pulp is often artifactually expanded with erythrocytes from barbiturate overdose at euthanasia. HE 4×. **b.** Unique to the pig, the capillaries are surrounded by phagocytes. HE 10×.

Thymus

Grossly, the thymus in the pig consists of two portions: a cervical and a thoracic portion. The cervical portion extends from the laryngeal area to the thoracic inlet in young animals and the thoracic portion is located at the thoracic inlet. Histologically, the outer cortex of darkly striating densely packed small lymphocytes surrounds a lighter-staining medulla (Figure 3.18). Similarly to humans, Hassall's corpuscles are present in the medulla of the thymus, and keratinization of Hassall's corpuscles is prominent in pigs (Haley 2017). B-cells may be observed in the porcine thymus as well (Rothkotter 2009).

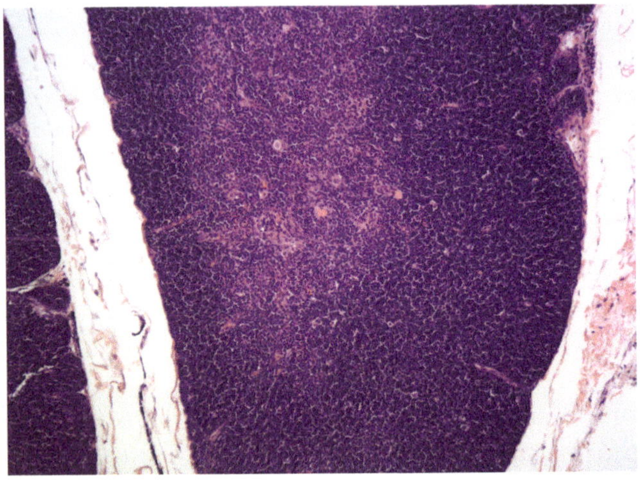

Figure 3.18 Thymus. Hassall's corpuscles are evident in the medulla of the porcine thymus. The cortex is composed of tightly packed lymphocytes. HE 10×.

HEPATOBILIARY SYSTEM

Liver

The liver of the pig has well-defined, prominent, well-organized lobules separated by fibrous septae (Figure 3.19). The amount of interlobular connective tissue does not differ based on sex in the pig, but there are differences among lobes, with the quadrate lobe having the most interlobular connective tissue (Mik et al. 2018). The lobules are composed primarily of cords of hepatocytes separated by hepatic sinusoids which contain phagocytic Kupffer cells. Most hepatocytes are mononuclear, but anywhere from 2–6% of the hepatocytes may be binucleated, with the binucleated cells more commonly observed in areas away from the periphery (Junatas et al. 2017).

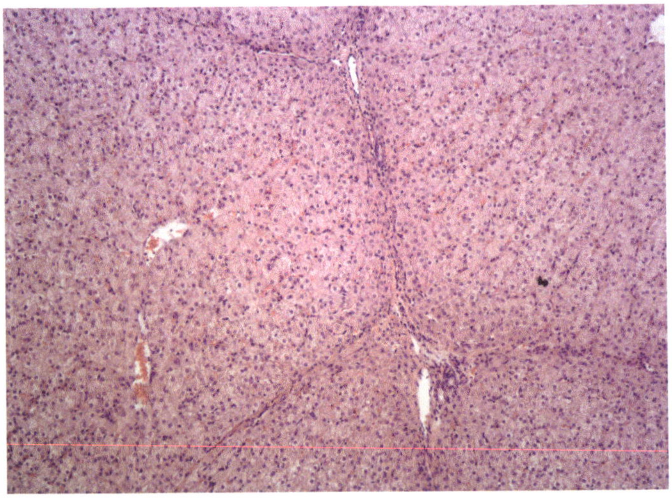

Figure 3.19 The liver has prominent lobes separated by fibrous septae. HE 10×.

Gallbladder

The gallbladder has simple or branched folds depending on the anatomical location of the epithelium (Figure 3.20). There are subepithelial mucous glands in the neck and in the portion of the body of the gallbladder fused with the liver (Prozorowska and Jackowiak 2015), which may be evident grossly as well as histologically. In Göttingen minipigs, one of the background findings that may be seen is chronic necrotizing cholecystitis and hypoplasia or aplasia of the gallbladder (Dincer and Skydsgaard 2012).

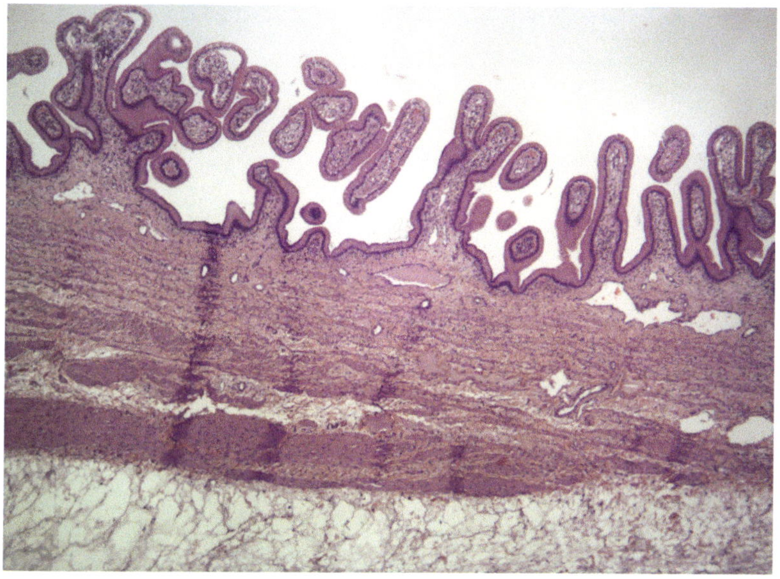

Figure 3.20 Gallbladder mucosa showing villi and loosely arranged submucosa. HE 4×.

INTEGUMENT

There are many similarities between human and porcine skin. There is similar hair follicle density along with increased keratinization compared with other species. Similarly to humans, pigs have approximately 11 follicles/cm^2, with a denser haircoat dorsally and a sparsely haired ventrum (Lamoureaux et al. 2017; Stricker-Krongrad et al. 2017). In the pig, the epidermal thickness appears to be consistent among locations, but there are regional differences in thickness of the stratum corneum (Khiao In et al. 2019). There are also geographical differences in thickness of the dermis, with the thickest layers on the dorsum and thinnest on the ventral abdomen at the midline, lateral thigh, axilla, and pinna, with no difference noted in appearance of dermal collagen (Lamoureaux et al. 2017).

In addition to the anatomical geographical differences, there are breed differences in pigmentation, dermal thickness, and number and size of sweat glands (Moyo, Gomes, and Erlwanger 2018). The integument of pigs has primarily apocrine sweat glands that extend into the subcutis (Khiao In et al. 2019), with the density greatest on the dorsal neck, lateral shoulder, and axilla, and decreasing in number extending caudally (Lamoureaux et al. 2017). The density of sebaceous glands is greatest on the dorsal neck, followed by the forehead and pinnae (Lamoureaux et al. 2017).

Due to the decreased dermal thickness in the area behind the ear, this site is commonly used as an area to administer subcutaneous injections (Figure 3.21). The skin on the flank (Figure 3.22) is considered the most humanlike, and the dermis of the ventral abdomen (Figure 3.23) is thinnest.

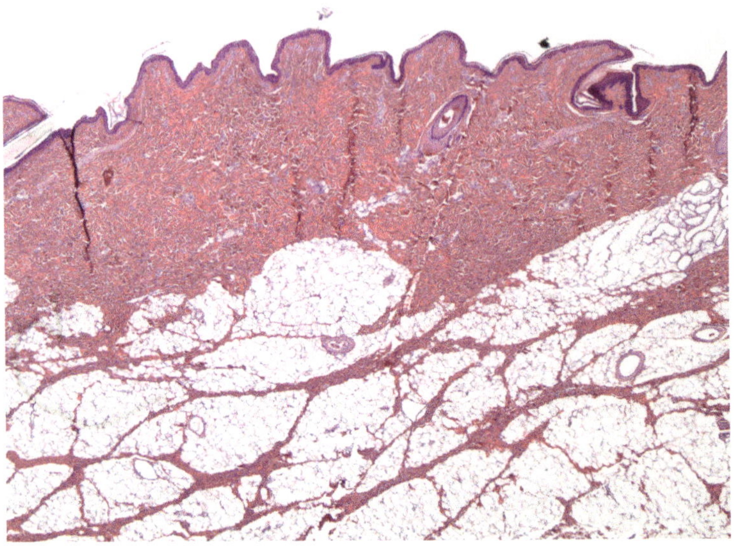

Figure 3.21 A thin epidermis covering the dermis and hypodermis of the porcine ear. HE 2×.

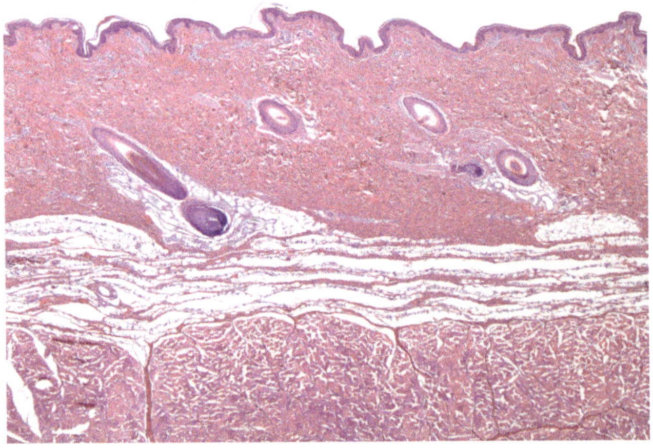

Figure 3.22 The hypodermis of the flank contains more collagen and connective tissue than on the ear and ventral abdomen. HE 2×.

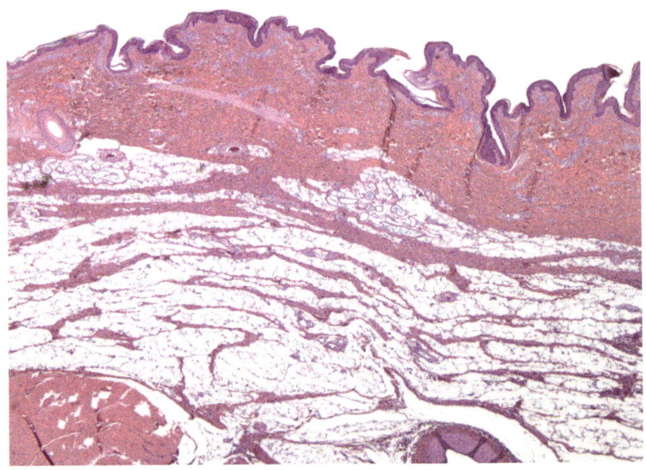

Figure 3.23 The dermis of the ventral abdomen is thinnest with a loosely arranged hypodermis. HE 2×.

MUSCULOSKELETAL SYSTEM

Bone

Porcine bone is similar to that of other mammals (Hillier and Bell 2007). There are two primary types of bone tissue: woven and lamellar. In addition, pigs also have plexiform bone in the cortical bone of the long bones, which is similar to lamellar bone, but has a more dense vascularization system that appears as a regularly spaced zone of empty spaces (vascular channels). Plexiform bone may appear throughout the entire long bone in immature pigs with no observable Haversian canals (Foote 1913). Cortical bone has several zones (periosteal, mesosteal, and endosteal) and cancellous bone is present on the interior of some bones.

There are sinuses within the skull of the pig that expand as the animal matures. As a young pig, there is only one layer of bone between the skin and the brain. As the animal ages, there will be two layers of bone that increase in thickness with age, separated by an air-filled sinus (Figures 3.24a and 3.24b) (Chang et al. 2012; Honig et al. 2002). This is relevant when trying to remove the brain for histology or perform cranial surgeries.

a

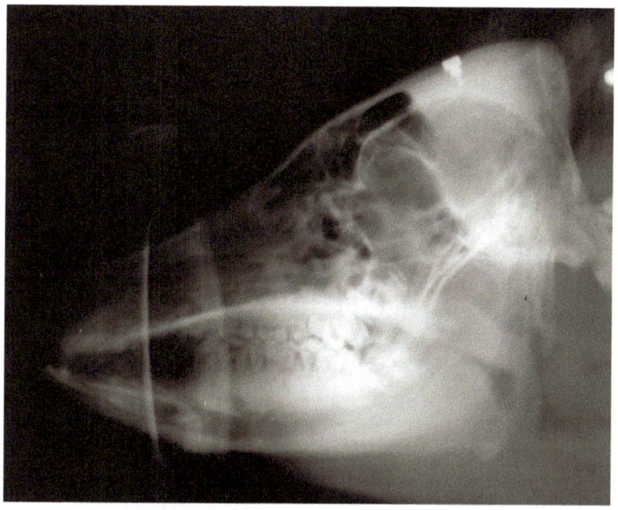

b

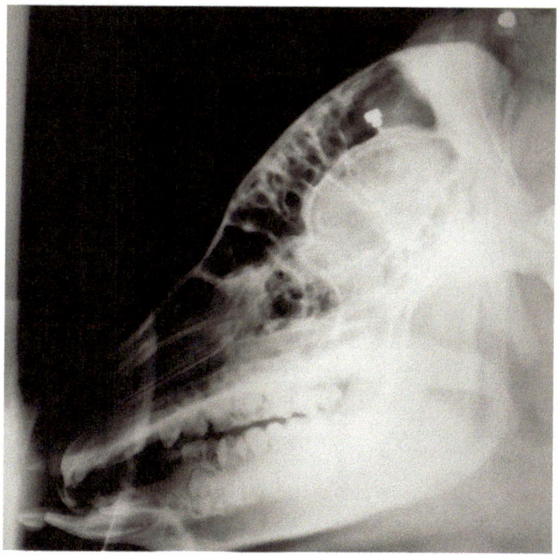

Figure 3.24 a. Radiograph of skull of 7-month-old Göttingen minipig with a metal implant in the skull. **b.** Radiograph of same animal at 14 months of age showing expansion of sinuses.

Skeletal Muscle

Skeletal muscle tissue in the pig is similar to that of other mammals (Figure 3.25). Several studies have shown that myoglobin increases in skeletal myocytes as pigs mature and age (Fitts, Nagle, and Cassens 1974; Morita et al. 1970). Myositis and myonecrosis has been reported in Göttingen minipigs as a background finding (Dincer and Skydsgaard 2012; McInnes and McKeag 2016).

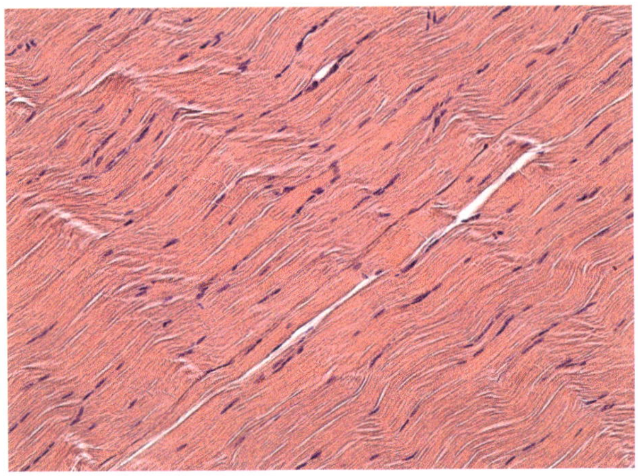

Figure 3.25 Skeletal muscle with numerous striations and nuclei to periphery. HE 20×.

Adipose

There are conflicting reports on the presence of brown adipose tissue in the pig, as brown fat has been defined as having uncoupling protein-1 (UCP-1), which has not yet been identified in the pig (Hou et al. 2017; Louveau et al. 2016; Mostyn et al. 2014).

Serous atrophy of the adipose tissue in the bone marrow of Göttingen minipigs is commonly reported (Dincer and Skydsgaard 2012; Skydsgaard et al. 2021).

NERVOUS SYSTEM

In the central nervous system of both pigs and humans, myelination increases with age. In pigs, the myelin increases from birth to the adult level at approximately 6 months of age (Fang et al. 2005). It is not just the thickness of the myelin sheath, but also the density of myelinated fibers that increases (Fang et al. 2005). The gray and white matter proportions are similar to those of humans (Conrad, Dilger, and Johnson 2012). There are several porcine brain atlases available: a stereotactic device for the pig (Marcilloux et al. 1989), a stereotactic atlas (Felix et al. 1999), and an atlas specific to the Göttingen minipig brain (Watanabe et al. 2001).

In order to access the brain of the pig, the skull must be broached using either a bone saw or rongeurs (Bjarkam et al. 2017; Helke 2015). The method used will depend on needed outcomes and age and breed of the pig, which influences the thickness of skull bones.

RENAL SYSTEM

The kidney is similar to that of the human in that it is multireniculate and multipapillate. The renal capsule has two distinct layers, one of dense collagen and elastic fibers, and the second composed of thin loose collagenous and reticular fibers interspersed with a few smooth muscle cells (Yadava 1955). Each kidney contains two to five arterial segments, with cranial and caudal segments being most common (Pereira-Sampaio et al. 2007).

Histologically, the glomeruli and the collecting system in the pig are similar to those of most other mammals (Figure 3.26), except that in the pig, the brush borders on the apical surface of proximal tubule epithelial cells are clustered, unlike the uniform arrangement found in other species (Figure 3.27) (Yadava 1955). In the Göttingen minipig, spontaneous glomerulonephritis has been reported and is more frequent in females than males (Helke et al. 2016; Vezzali et al. 2011).

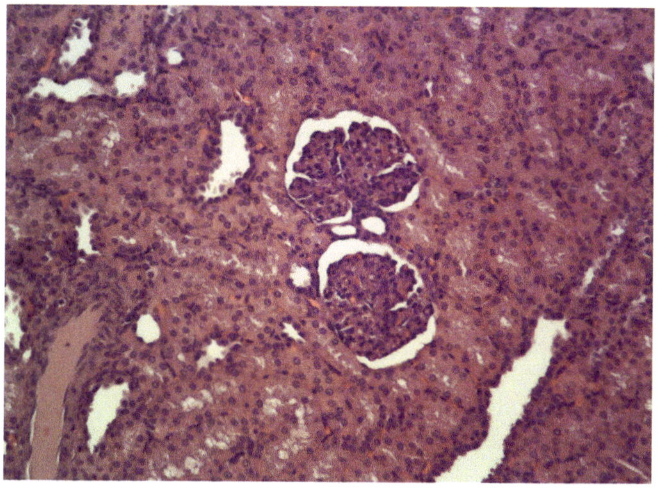

Figure 3.26 Kidney. Glomerular tufts surrounded by renal tubules. HE 20×.

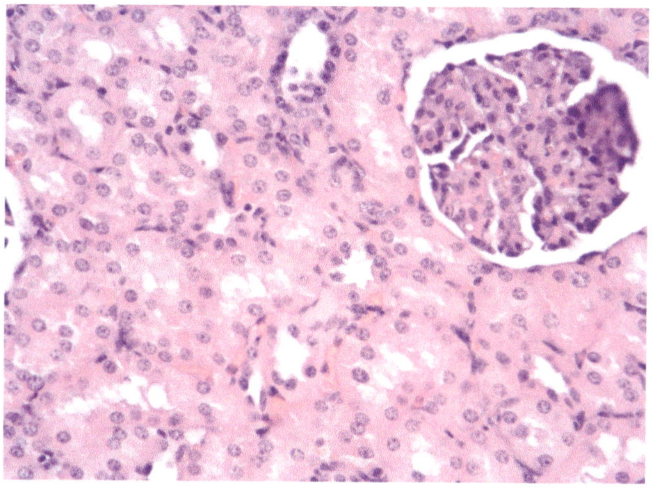

Figure 3.27 Kidney. The epithelial cells of the proximal tubules have clumped apical villi on the cell surface. HE 40×.

REPRODUCTIVE TRACT

Female

Histology of the ovaries and uterus is used to stage the estrous cycle in minipigs (de Rijk et al. 2014). Tertiary follicles in the ovary are most reliably used for staging along with Graafian follicles and corpora lutea; see de Rijk's article for further information and descriptions (de Rijk et al. 2014).

The uterus is composed of three zones: the luminal epithelium, the glandular layer, and the basal glandular layer, all of which are surrounded by the myometrium (Figure 3.28). The porcine cervix has a thick muscular wall with abundant elastic fibers and is quite elongated, up to several inches in length (Lorenzen et al. 2015).

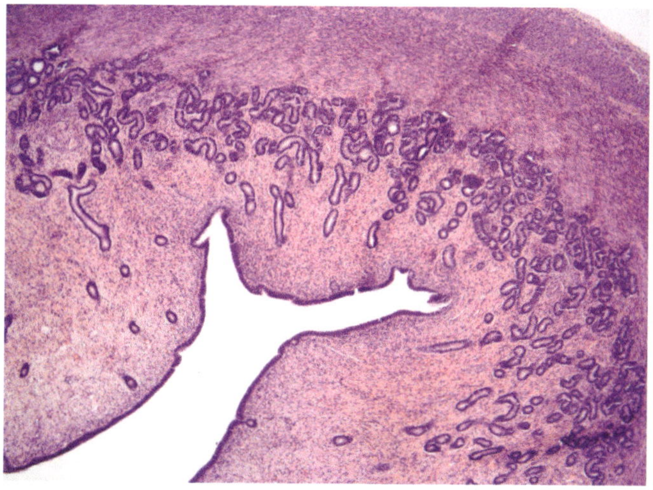

Figure 3.28 Uterus. The luminal epithelial cells overlie the mucosa formed of uterine glands. This layer is surrounded by two layers of muscularis. HE 4×.

Male

It is recommended that the testes and epididymides be fixed in modified Davidson's fixative to prevent shrinkage and provide better histological morphology (Albl et al. 2016).

The cell types present in the porcine testis are not especially different from those of other species (Taberner et al. 2016). Brightly eosinophilic Leydig cells are more prominent in the porcine testis compared to other species. During the first 6 months of age, the Leydig cells increase in volume with age (Figure 3.29) (Penchev 2011), with up to one-third of the testicular volume composed of Leydig cells. The Göttingen minipigs are prone to testicular hypoplasia/atrophy as a background finding (Helke et al. 2016). Boars have a compact prostate gland similar to that of humans (Figure 3.30).

Spermatozoa with or without multinucleated germ cells are found within the caput (distal portion) and cauda epididymis (Kangawa et al. 2016).

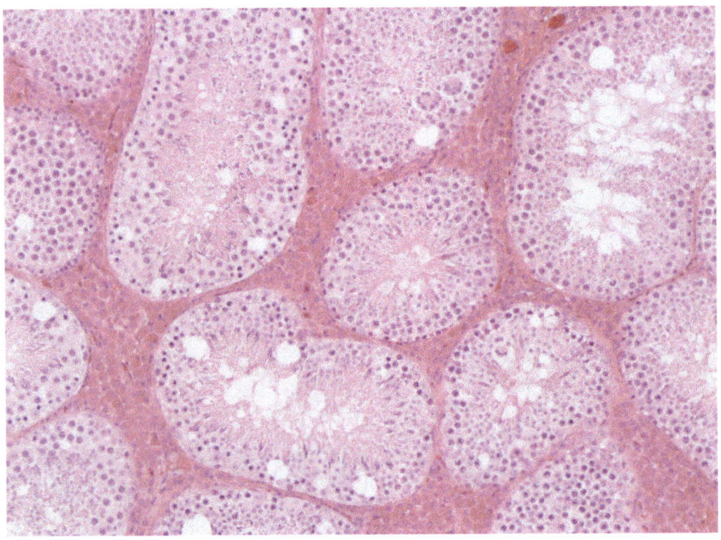

Figure 3.29 The Leydig cells are especially prominent in the porcine testis. Seminiferous tubules are similar to those of other species. HE 10×.

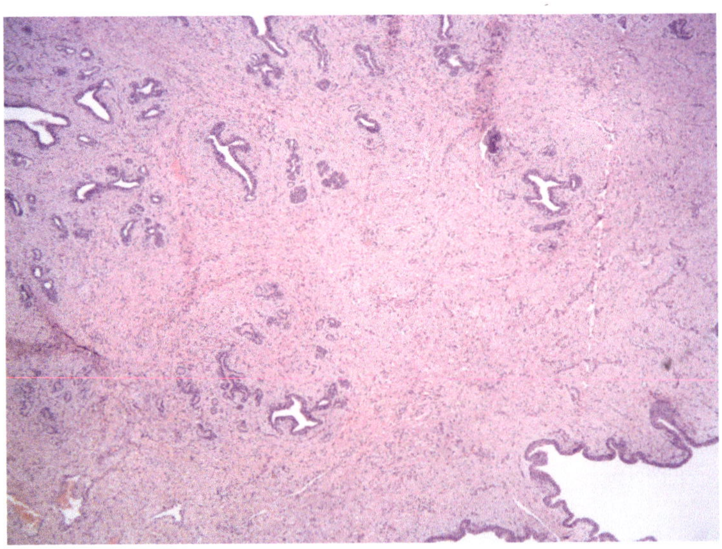

Figure 3.30 The porcine prostate has dense connective tissue with scattered glands throughout the tissue. The urethra is present in the lower right corner. HE 4×.

RESPIRATORY SYSTEM

Grossly, the lung has thickened pleura with fibrous septae between lobes, which is also evident histologically (Figure 3.31). Within the lung, the bronchioles are lined by both non-ciliated and ciliated cells (Baskerville 1970). Rarely, neuroendocrine cells may be found in the pig lung in the epithelium of the intrapulmonary airways, often at the bifurcation of airways (DiAugustine and Sonstegard 1984).

Bronchioli in the pig are composed of epithelium surrounded by smooth muscle and connective tissue. Alveoli are lined by type I pneumocytes, but type II pneumocytes are abundant as well (Baskerville 1970). Bronchiole-associated lymphoid tissue (BALT) may be found normally at the bifurcation of larger airways (Haley 2017). Intra-alveolar macrophages are commonly seen in the alveoli and are considered normal (Figure 3.32).

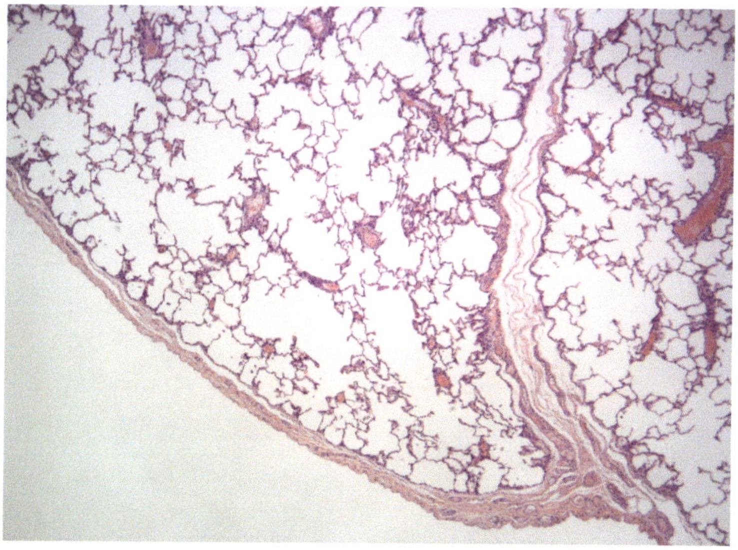

Figure 3.31 Lung. The pleura is prominent in the porcine lung and extends between lobes. HE 4×.

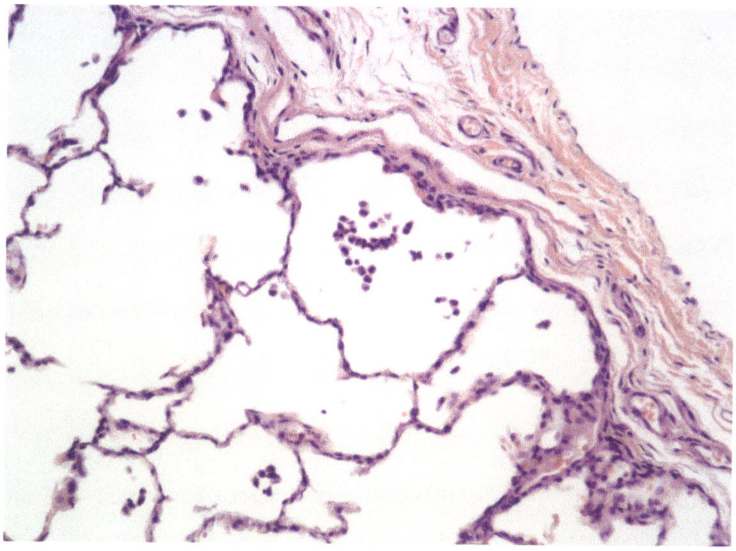

Figure 3.32 Lung. Macrophages can be found within the alveoli in the lungs of normal pigs. HE 20×.

Since pigs are often used as surgical models and for testing of medical devices, animals may be placed on a ventilator for the surgical procedure. If the animal is on a ventilator for any reason (surgical, device studies), there are changes in the lung to be aware of which differ based on how long the animals is on the ventilator (Fremuth et al. 2014). Another factor for consideration when performing histological examination of the lung is the rooting behavior of pigs. They often inhale food or bedding particles into nasal passages and into airways (Skydsgaard et al. 2021). These foreign particles may be observed histologically, or there may be accumulations of macrophages or giant cells containing foreign material.

In the lung, blood vessels of similar size have noticeably thicker walls in pigs compared to other species (Hedenstierna et al. 2000). The pulmonary microvasculature also contains pulmonary intravascular macrophages (PIMs) (Figure 3.33). These cells resemble the Kupffer cells of the liver and help with clearance of blood-borne pathogens and particulates (Winkler 1989). The PIMs also increase in number with maturity (Winkler 1989).

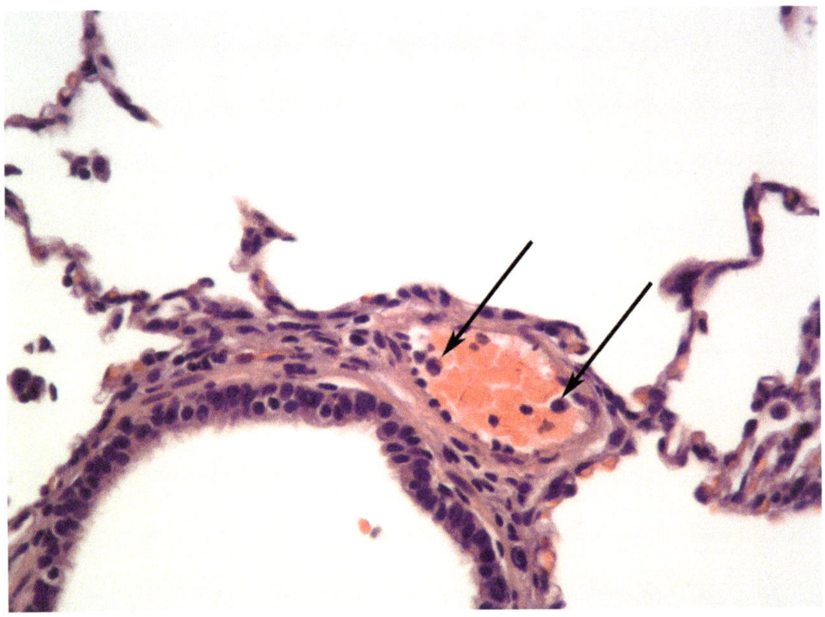

Figure 3.33 Lung, vessel. There are intravascular macrophages (arrows) found normally in the porcine lung. HE 40×.

SPECIAL SENSES

Eye

There are sex differences in retinal layer thickness, specifically the ganglion cell layer, inner plexiform layer, and outer plexiform layer (Shrader and Mowry 2019). Eyes and structures within the eyes do not reach maturity until 6 months of age in either Göttingen or farm breeds (Vrolyk et al. 2020); other breeds were not examined in this study.

CONCLUSIONS

There are several histological differences between the pig and other species. Knowledge of these species-specific findings will help in analysis of porcine tissue for diagnostic and research purposes.

REFERENCES

Akaydin, Y., and M. Kabak. 2006. First description and morphology of haemal nodes in piglets (*Sus scrofa domestica*). *Acta Vet Hung* 54 (2):135–42.

Albl, B., S. Haesner, C. Braun-Reichhart, et al. 2016. Tissue sampling guides for porcine biomedical models. *Toxicol Pathol* 44 (3):414–20.

Bal, H. S., and N. G. Ghoshal. 1972. Histomorphology of the torus pyloricus of the domestic pig (*Sus scrofa domestica*). *Zentralbl Veterinarmed C* 1 (4):289–98.

Baskerville, A. 1970. Ultrastructural studies of the normal pulmonary tissue of the pig. *Res Vet Sci* 11 (2):150–5.

Bjarkam, C. R., D. Orlowski, L. Tvilling, J. Bech, A. N. Glud, and J. H. Sorensen. 2017. Exposure of the pig CNS for histological analysis: A manual for decapitation, skull opening, and brain removal. *J Vis Exp* (122):55511.

Boes, K. M., and A. C. Durham. 2016. Bone marrow, blood cells, and the lymphoid/lymphatic system. In *Pathologic Basis of Veterinary Disease*, edited by J. F. Zachary. St. Louis, MO: Elsevier.

Bornstein, S. R., M. Ehrhart-Bornstein, H. Usadel, M. Bockmann, and W. A. Scherbaum. 1991. Morphological evidence for a close interaction of chromaffin cells with cortical cells within the adrenal gland. *Cell Tissue Res* 265 (1):1–9.

Chang, E. H., A. A. Pezzulo, D. K. Meyerholz, et al. 2012. Sinus hypoplasia precedes sinus infection in a porcine model of cystic fibrosis. *Laryngoscope* 122 (9):1898–905.

Conrad, M. S., R. N. Dilger, and R. W. Johnson. 2012. Brain growth of the domestic pig (*Sus scrofa*) from 2 to 24 weeks of age: A longitudinal MRI study. *Dev Neurosci* 34 (4):291–8.

de Rijk, E., H. van den Brink, J. Lensen, A. Lambregts, H. Lorentsen, and B. Peter. 2014. Estrous cycle-dependent morphology in the reproductive organs of the female Göttingen minipig. *Toxicol Pathol* 42 (8):1197–211.

DiAugustine, R. P., and K. S. Sonstegard. 1984. Neuroendocrinelike (small granule) epithelial cells of the lung. *Environ Health Perspect* 55:271–95.

Dincer, Z., V. Piccicuto, U. J. Walker, A. Mahl, and S. McKeag. 2018. Spontaneous and drug-induced arteritis/polyarteritis in the Göttingen minipig – Review. *Toxicol Pathol* 46 (2):121–30.

Dincer, Z., and M. Skydsgaard. 2012. Spontaneous/background pathology of Göttingen minipig. In *The Minipig in Biomedical Research*, edited by P. McAnulty, A. D. Dayan, N.-C. Ganderup, and K. Hastings. Boca Raton, FL: CRC Press.

Elling, F., E. Hasselager, and C. Friis. 1977. Perfusion fixation of kidneys in adult pigs for electron microscopy. *Acta Anat (Basel)* 98 (3):340–2.

Fang, M., J. Li, X. Gong, et al. 2005. Myelination of the pig's brain: A correlated MRI and histological study. *Neurosignals* 14 (3):102–8.

Felix, B., M. E. Leger, D. Albe-Fessard, J. C. Marcilloux, O. Rampin, and J. P. Laplace. 1999. Stereotaxic atlas of the pig brain. *Brain Res Bull* 49 (1-2):1–137.

Fitts, R. H., F. J. Nagle, and R. G. Cassens. 1974. The adaptation of myoglobin with age and training and its relationship to the three fiber types of skeletal muscle in miniature pig. *Eur J Appl Physiol Occup Physiol* 33 (4):275–84.

Foote, J. S. 1913. The Comparative Histology of the Femur. Washington, DC: Smithsonian Institution.

Fremuth, J., J. Kobr, K. Pizingerova, et al. 2014. Healthy lung tissue response to mechanical ventilation in an experimental porcine model. *In Vivo* 28 (5):803–9.

Fujioka, T. 1956. Studies on the histology of the adrenal cortex in the pig. *Nihon Chikusan Gakkaiho* 27 (1):19–30.

Gonzalez, L. M., A. J. Moeser, and A. T. Blikslager. 2015. Porcine models of digestive disease: The future of large animal translational research. *Transl Res* 166 (1):12–27.

Gonzalez, L. M., I. Williamson, J. A. Piedrahita, A. T. Blikslager, and S. T. Magness. 2013. Cell lineage identification and stem cell culture in a porcine model for the study of intestinal epithelial regeneration. *PLOS ONE* 8 (6):e66465.

Green, W. B., K. Eaton, and S. Krakowka. 1997. Porcine gastric mucosa-associated lymphoid tissue (MALT): Stimulation by colonization with the gastric bacterial pathogen, *Helicobacter pylori*. *Vet Immunol Immunopathol* 56 (1-2):119–31.

Haley, P. J. 2017. The lymphoid system: A review of species differences. *J Toxicol Pathol* 30 (2):111–23.

Hedemann, M. S., E. Kristiansen, and G. Brunsgaard. 2002. Morphology of the large intestine of the pig: Haustra versus taenia. *Ann Anat* 184 (4):401–3.

Hedenstierna, G., M. Hammond, O. Mathieu-Costello, and P. D. Wagner. 2000. Functional lung unit in the pig. *Respir Physiol* 120 (2):139–49.

Helke, K. L. 2015. Necropsy on research swine. In *Swine in the Laboratory: Surgery, Anesthesia, Imaging, and Experimental Techniques*, edited by M. M. Swindle, and A. C. Smith. Boca Raton, FL: CRC Press.

Helke, K. L., K. N. Nelson, A. M. Sargeant, et al. 2016. Background pathological changes in minipigs: A comparison of the incidence and nature among different breeds and populations of minipigs. *Toxicol Pathol* 44 (3):325–37.

Hillier, M. L., and L. S. Bell. 2007. Differentiating human bone from animal bone: A review of histological methods. *J Forensic Sci* 52 (2):249–63.

Honig, J. F., H. A. Merten, R. Schutte, U. A. Grohmann, and A. Cassisis. 2002. Experimental study of the frontal sinus development on Goettingen miniature pigs. *J Craniofac Surg* 13 (3):418–26.

Hou, L., J. Shi, L. Cao, G. Xu, C. Hu, and C. Wang. 2017. Pig has no uncoupling protein 1. *Biochem Biophys Res Commun* 487 (4):795–800.

Hunt, A. C. 1968. Micro-anatomy of the lymph nodes of the pig. *Br J Exp Pathol* 49 (4):338–9.

Igbokwe, C. O., and D. N. Ezeasor. 2015. Histological and immunohistochemical changes of the thyroid gland during the foetal and post-natal period of development in indigenous Large White crossbred pigs. *Bulgarian J Vet Med* 18 (4):313–24.

Junatas, K. L., Z. Tonar, T. Kubikova, et al. 2017. Stereological analysis of size and density of hepatocytes in the porcine liver. *J Anat* 230 (4):575–88.

Kangawa, A., M. Otake, S. Enya, T. Yoshida, and M. Shibata. 2016. Histological development of male reproductive organs in microminipigs. *Toxicol Pathol* 44 (8):1105–22.

Kararli, T. T. 1995. Comparison of the gastrointestinal anatomy, physiology, and biochemistry of humans and commonly used laboratory animals. *Biopharm Drug Dispos* 16 (5):351–80.

Khiao In, M., K. C. Richardson, A. Loewa, S. Hedtrich, S. Kaessmeyer, and J. Plendl. 2019. Histological and functional comparisons of four anatomical regions of porcine skin with human abdominal skin. *Anat Histol Embryol* 48 (3):207–17.

Lamoureaux, J., D. Snider, M. Ashley, and K. N. Nelson. 2017. Variance in regional skin histoanatomy in Göttingen minipigs. In *Society of Toxicologic Pathology Annual Symposium*. Montreal, QC, Canada.

Lanning, L. L., D. M. Creasy, R. E. Chapin, et al. 2002. Recommended approaches for the evaluation of testicular and epididymal toxicity. *Toxicol Pathol* 30 (4):507–20.

Li, Y., Z. Song, K. A. Kerr, and A. J. Moeser. 2017. Chronic social stress in pigs impairs intestinal barrier and nutrient transporter function, and alters neuro-immune mediator and receptor expression. *PLoS One* 12 (2):e0171617.

Lorenzen, E., F. Follmann, G. Jungersen, and J. S. Agerholm. 2015. A review of the human vs. porcine female genital tract and associated immune system in the perspective of using minipigs as a model of human genital *Chlamydia* infection. *Vet Res* 46:116.

Louveau, I., M. H. Perruchot, M. Bonnet, and F. Gondret. 2016. Invited review: Pre- and postnatal adipose tissue development in farm animals: From stem cells to adipocyte physiology. *Animal* 10 (11):1839–47.

Marcilloux, J. C., O. Rampin, M. B. Felix, J. P. Laplace, and D. Albe-Fessard. 1989. A stereotaxic apparatus for the study of the central nervous structures in the pig. *Brain Res Bull* 22 (3):591–7.

McInnes, E. F., and S. McKeag. 2016. A brief review of infrequent spontaneous findings, peculiar anatomical microscopic features, and potential artifacts in Göttingen minipigs. *Toxicol Pathol* 44 (3):338–45.

Mekbungwan, A., N. Thongwittaya, and K. E. Yamauchi. 2004. Digestibility of soybean and pigeon pea seed meals and morphological intestinal alterations in pigs. *J Vet Med Sci* 66 (6):627–33.

Mik, P., Z. Tonar, A. Maleckova, et al. 2018. Distribution of connective tissue in the male and female porcine liver: Histological mapping and recommendations for sampling. *J Comp Pathol* 162:1–13.

Morita, S., C. C. Cooper, R. G. Cassens, L. L. Kastenschmidt, and E. J. Briskey. 1970. A histological study of myoglobin in developing muscle of the pig. *J Anim Sci* 31 (4):664–70.

Mostyn, A., L. Attig, T. Larcher, et al. 2014. UCP1 is present in porcine adipose tissue and is responsive to postnatal leptin. *J Endocrinol* 223 (1):M31–8.

Moyo, D., M. Gomes, and K. H. Erlwanger. 2018. Comparison of the histology of the skin of the Windsnyer, Kolbroek and Large White pigs. *J S Afr Vet Assoc* 89 (0):e1-e10.

Mulligan-Kehoe, M. J. 2010. The vasa vasorum in diseased and nondiseased arteries. *Am J Physiol Heart Circ Physiol* 298 (2):H295–305.

Musigazi, G. U., S. De Vleeschauwer, R. Sciot, E. Verbeken, and B. Depreitere. 2018. Brain perfusion fixation in male pigs using a safer closed system. *Lab Anim* 52 (4):413–17.

Oster, M., J. Keiler, M. Schulze, H. Reyer, A. Wree, and K. Wimmers. 2018. Fast and reliable dissection of porcine parathyroid glands – A protocol for molecular and histological analyses. *Ann Anat* 219:76–81.

Pabst, R., M. Geist, H. J. Rothkotter, and F. J. Fritz. 1988. Postnatal development and lymphocyte production of jejunal and ileal Peyer's patches in normal and gnotobiotic pigs. *Immunology* 64 (3):539–44.

Penchev, G. 2011. Postnatal development of interstitial endocrine (Leydig) cells in pig's testes – Electron microscopy study. *Bulgarian J Vet Med* 14 (3):131–6.

Pereira-Sampaio, M., L. A. Favorito, R. Henry, and F. J. Sampaio. 2007. Proportional analysis of pig kidney arterial segments: Differences from the human kidney. *J Endourol* 21 (7):784–8.

Prozorowska, E., and H. Jackowiak. 2015. Light microscopy and scanning electron microscopy study on microstructure of gallbladder mucosa in pig. *Microsc Res Tech* 78 (3):220–9.

RENI, Global Open. 2020. *Global Open RENI: The standard reference for nomenclature and diagnostic criteria in toxicologic pathology*, Oct 2020 [cited 16 Oct 2020]. Available from https://www.goreni.org/gr3_index.php.

Rinke, M. 1997. How clean is a mini-pig? – Impressions and suggestions of a pathologist working in the field of toxicology. *Pharmacol Toxicol* 80 Suppl 2:16–22.

Rothkotter, H. J. 2009. Anatomical particularities of the porcine immune system – A physician's view. *Dev Comp Immunol* 33 (3):267–72.

Sedmera, D., and R. G. Gourdie. 2014. Why do we have Purkinje fibers deep in our heart? *Physiol Res* 63 (Suppl 1):S9–18.

Shrader, S. M., and R. N. Mowry. 2019. Histomorphometric evaluation of the Göttingen minipig eye. *Vet Ophthalmol* 22 (6):872–8.

Shringi, N., R. Mathur, V. Kumar, K. Rohlan, and S. Ganguly. 2018. Histological studies on the spleen of Large White Yorkshire pig (*Sus scrofa*). *J Entemol Zool Stud* 6 (1):1142–4.

Simões, M., D. B. de Souza, C. B. Gallo, M. A. Pereira-Sampaio, W. S. Costa, and F. J. Sampaio. 2016. Histomorphometric comparison of the human, swine, and ovine collecting systems. *Anat Rec (Hoboken)* 299 (7):967–72.

Skydsgaard, M., Z. Dincer, W. Haschek, et al. 2021. International Harmonization of Nomenclature and Diagnostic Criteria (INHAND): Nonproliferative and proliferative lesions of the minipig. *Toxicol Pathol* 49 (1).

Soshin, T., H. Takai, C. Kato, et al. 2010. A method for sampling and tissue preparation of the parathyroid glands in miniature pigs for toxicity studies. *J Toxicol Sci* 35 (2):235–8.

Stricker-Krongrad, A., C. R. Shoemake, J. Liu, D. Brocksmith, and G. Bouchard. 2017. The importance of minipigs in dermal safety assessment: An overview. *Cutan Ocul Toxicol* 36 (2):105–13.

Taberner, E., N. Navratil, B. Jasmin, M. Salerno, B. Grambo, and G. C. Althouse. 2016. Pubertal age based on testicular and epididymal histology in Göttingen minipigs. *Theriogenology* 86 (9):2091–5.

Tay, H., T. Vandecasteele, and W. Van den Broeck. 2017. Identification of telocytes in the porcine heart. *Anat Histol Embryol* 46 (6):519–27.

Tsuchitani, M., J. Sato, and H. Kokoshima. 2016. A comparison of the anatomical structure of the pancreas in experimental animals. *J Toxicol Pathol* 29 (3):147–54.

Udoumoh, A. F., D. N. Ezeasor, and U. M. Igwebuike. 2016. Distribution of porcine haemal nodes and morphological variations of the haemal nodes of cattle and West African dwarf goats. *Not Sci Biol* 8 (4):422–7.

Vezzali, E., R. A. Manno, D. Salerno, G. Oberto, A. Nyska, and Y. Ramot. 2011. Spontaneous glomerulonephritis in Göttingen minipigs. *Toxicol Pathol* 39 (4):700–5.

Vrolyk, V., M. J. Desmarais, D. Lambert, J. Haruna, and M. O. Benoit-Biancamano. 2020. Neonatal and juvenile ocular development in Göttingen minipigs and domestic pigs: A histomorphological and immunohistochemical study. *Vet Pathol* 57 (6):889–914.

Watanabe, H., F. Andersen, C. Z. Simonsen, et al. 2001. MR-based statistical atlas of the Göttingen minipig brain. *Neuroimage* 14 (5):1089–96.

Winkler, G. C. 1989. Review of the significance of pulmonary intravascular macrophages with respect to animal species and age. *Exp Cell Biol* 57 (6):281–6.

Yadava, R. P. 1955. Comparative Histology of the Kidney of Domestic Animals. Department of Anatomy, Michigan State University of Agriculture and Applied Science, Ann Arbor, MI.

Yen, J. 2001. Anatomy of the digestive system and nutritional physiology. In *Swine Nutrition*, edited by A. J. Lewis, and L. L. Southern. Boca Raton, FL: CRC Press.

Index

Q

R

S

T

U

V